Publications of the Midwest Institute of Geosciences and Engineering Past and Present
VOLUME 1

Compiled by: Steven D.J. Baumann

Cover photo was taken by Steven D.J. Baumann on May 6, 2019 at 18:22 UTC. The outcrop is an ARTERITE MIGMATITE, taken on the north shore of Lake Superior along Trans-Canada 17

Spring 2021

<u>**VOLUME 1 LETTER OF INTENT**</u>

This is the first compilation volume of publications done by the Midwest Institute of Geosciences and Engineering (MIGE) that have not been publicly released. All of us at MIGE love to publicly post our work, but we need to fund raise in order to keep working. So we have collectively decided on March 25, 2021, that we would begin taking older unpublished works and assembling them into volumes such as this. There is no set schedule. Things like large maps, will continue to be available for free. We hope you understand why we are doing this. We figure if we are going to fundraise, you might as well as get something in return.

These have already been written and were going to be put on the website. Based on the recent decision, some have been reformatted to fit the page and obvious errors were corrected. Beyond that, the publications are "as written".

Please enjoy!

Sincerely,

Steven D.J. Baumann, P.G.: Founder and President

Table of Contents

Written by: *Steven D.J. Baumann*

G-072015-1B

Updated: March 23, 2021 (version 2)

LOCATION:

Ableman's Gorge is located 0.39 miles north of the intersection of State Route 136 (East Broadway Street) and County Road DD, along the west side of State Route 136.

Parking to the gorge is located at GPS: 43.48342° −89.91753°, just north of the tan colored aluminum building used for the loading of spring water. The South Wall at Ableman's Gorge is located at GPS: 43.48355° −89.91864°,

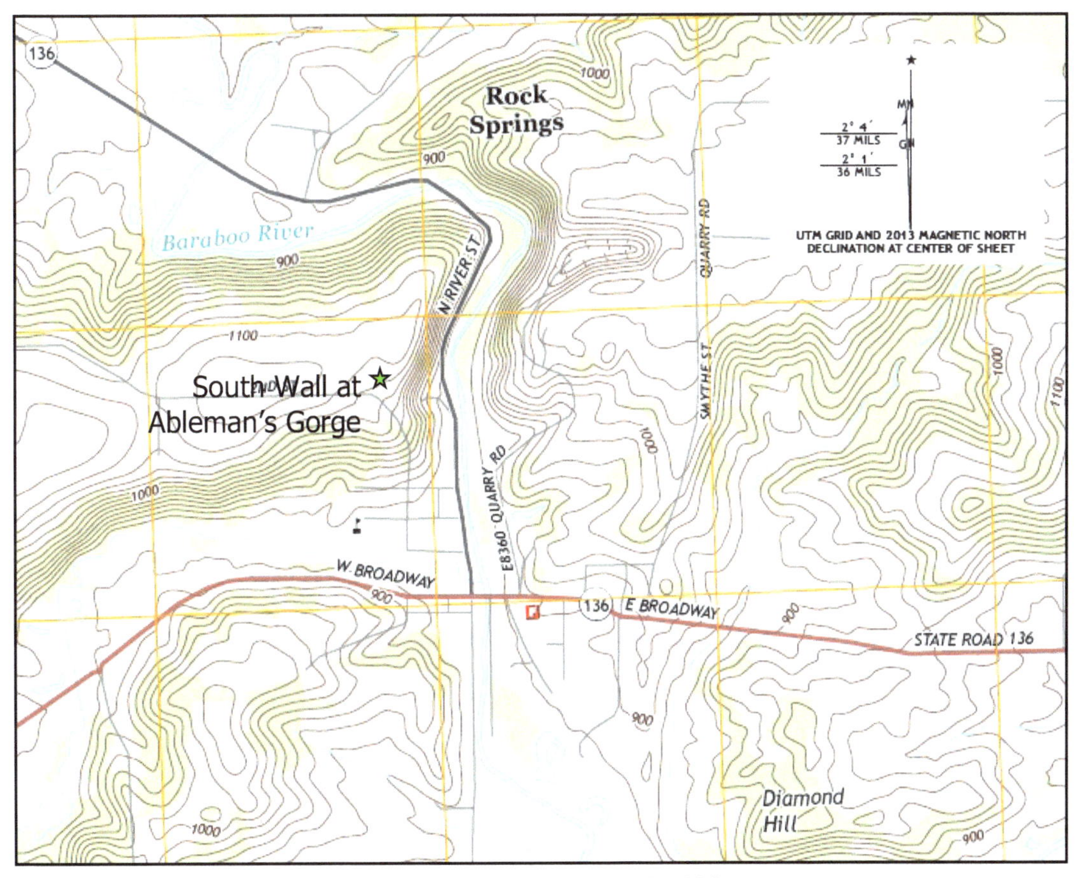

SCALE 1:24 000

QUADRANGLE LOCATION

Reedsburg West	Reedsburg East	Wisconsin Dells South
Loganville	Rock Springs	North Freedom
Plain	Black Hawk	Sauk Prairie

ADJOINING 7.5′ QUADRANGLES

ROCK SPRINGS, WI

2013

CONTOUR INTERVAL 20 FEET
NORTH AMERICAN VERTICAL DATUM OF 1988

This map was produced to conform with the
National Geospatial Program US Topo Product Standard, 2011.
A metadata file associated with this product is draft version 0.6.12

HISTORICAL BACKGROUND:

Ableman's Gorge was originally a natural cliff face in the Baraboo Quartzite. In 1851, a man by the name of S.V.R. Ableman (a Colonel in the U.S. Army) decided to settle the area in what is now called the town of Rock Springs. The town that Col. Ableman would settle would bear his name until 1875, when the town's name was changed to Rock Springs, which reverted back to Ableman in 1879, and finally back to Rock Springs in 1947.

The Gorge itself opened as a stone quarry in 1870, in order to provide ballast for the Chicago and Northwestern Railway. The section of this railway that extends through the town still exists and is active today. I could not find any information on when Ableman's quarry ceased operations. However, E.O. Ulrich visited the quarry in 1916. It was apparently still active at that time.

GEOLOGICAL BACKGROUND:

The geologic history of Ableman's Gorge begins about 1.72 billion years (Ga) ago, when the purple quartzite was deposited as a medium to coarse grained quartz rich sand in an ancient shallow sea. The sand became buried and metamorphosed into quartzite during a mountain building event (orogeny) that occurred at 1.62Ga. It was during this orogeny that the Baraboo Quartzite was turned vertical in the immediate area of Rock Springs.

Roughly 1.5Ga later, the mountains would begin to erode and the quartzite stood as islands during the invasion of the Cambrian seas. The quartzite would then become buried by sediments once again until about 350 million years (Ma) ago when the seas retreated and erosion once again took over.

The gorge itself was likely carved out slowly, beginning 2Ma years ago. Floodwaters cut through the valley

THE FALL OF THE SOUTH WALL:

The south wall of Ableman's Gorge exposes a long wall of ripple marks that formed as the Baraboo was being deposited as sand 1.72 to 1.65Ga ago. The ripple marks were turned vertical during metamorphism, about 1.6 to 1.5Ga. These iconic ripples were especially well preserved on the east end of south wall. Van Hise himself looked at these ripple marks when he figured out the mechanics of mountain building during the early 20th century. Since then, generations of geologists have visited this outcrop on their structural and sedimentary geology class trips.

The fate of the ripple marks was sealed from the moment they were exposed. The vertically tilted beds that bear the ripple marks are a weak zone within the quartzite. Sometime during the winter of 2014-2015 or the spring of 2015, these iconic ripples lost their battle with nature and came tumbling down. Their destruction can be attributed to a natural erosional process known as "frost wedging". This is a physical form of weathering in which water gets into small cracks in the rock (in this case along the bedding plane), freezes during the winter, the ice expands, thus slowly widening the cracks, or fractures. Eventually the fractures open so much that the rock loses stability, and in the case of the south wall, it catastrophically failed.

The exact date of the collapse isn't known. During the Illinois State University's 2014 structural geology field trip in autumn, the wall was still intact. When I visited the wall in late July 2015, the wall had failed and the iconic ripples had been strewn about as talus. New , less developed, ripples were exposed. Thus beginning the process all over again.

The iconic ripples were not the first. Before the collapse, there was a talus pile at the base, indicating that another set of ripples had collapsed from this wall sometime after the floods from the last ice age ceased,

FROST WEDGING ON THE SOUTH WALL:

<u>Cross Section of the South Wall</u>

<u>Initial phase: Unweathered rock</u>

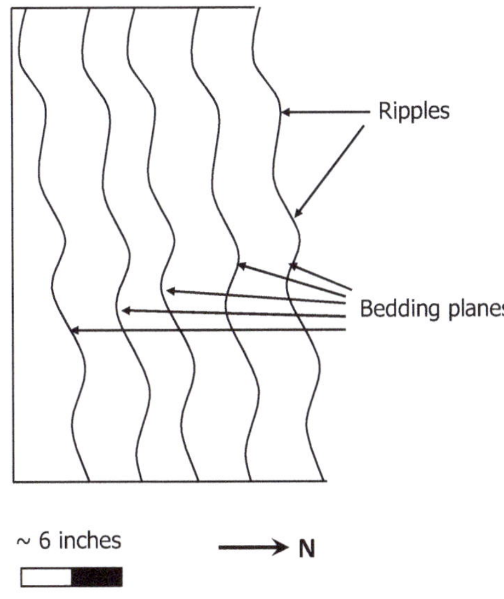

<u>Primary phase: water erosion</u>

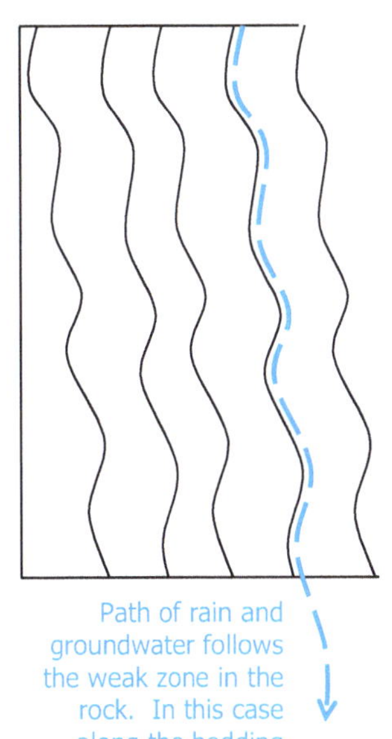

<u>Cross Section of the South Wall</u>

<u>Secondary phase: ice erosion</u>

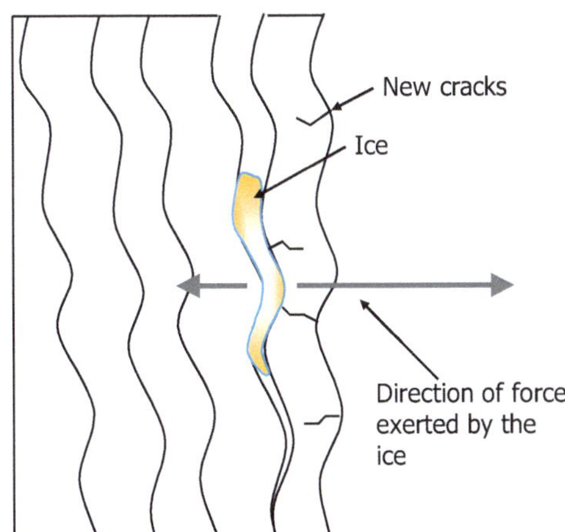

Over many years, the ice in the cracks freezes, slightly pushing on the rocks, gradually weakening the rock face.

<u>Final phase: rock failure</u>

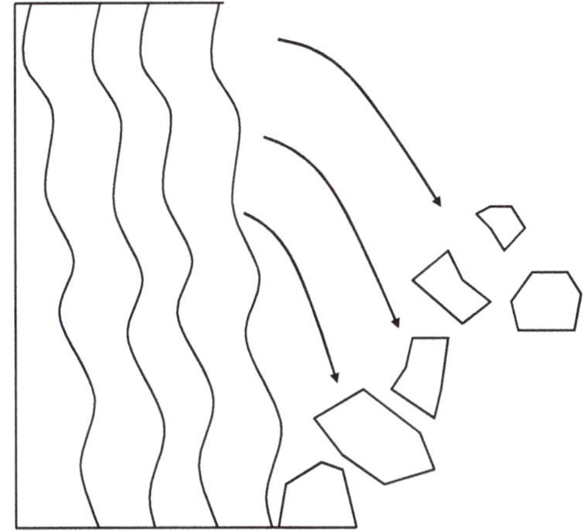

Eventually the rock weakens so much that it catastrophically fails, exposing a new surface, and the process starts over.

Illinois State University's structural geology field trip, looking east-southeast.

Photo was taken on October 8, 2006 by Steven D.J. Baumann.

South wall before the collapse of the iconic ripple marks, looking south and up. Notice the small ridges covering the surface. Those are the ripple marks.

Photo was taken on June 25, 2013 by Steven D.J. Baumann.

South wall showing the iconic ripples before the collapse, looking south. Chad Pilcher for scale.

Photo was taken on June 25, 2013 by Steven D.J. Baumann.

South wall after the collapse, looking south. Steven D.J. Baumann for scale.

Photo was taken on July 25, 2015 Rachel Hannaford.

Person is standing in almost the exact same spot in both photos.

Zoom out of photo number 4.

Photo was taken on July 25, 2015 by Steven D.J. Baumann.

After the collapse, looking east-southeast.

Photo was taken on July 25, 2015 by Steven D.J. Baumann.

REFERENCES:

Baumann, S.D.J., 2011, *Surficial Geologic Map of the Upper Narrows near Rock Springs, Sauk County, Wisconsin, U.S.A.*, Midwest Institute of Geosciences and Engineering, M-092011-3A

Baumann, S.D.J., Krol, Cory, A.B., Krol, M.M., and Piispa, E.J., 2017, *Precambrian Geologic Events in the Mid-Continent of North America*, Midwest Institute of Geosciences and Engineering, G-012011-1J

Dalziel, I.W.D., Dott, R.H., 1970, *Geology of the Baraboo District Wisconsin*, Geological and Natural History Survey, Wisconsin, Information Circular Number 14

Sauk County Historical society's website
http://www.saukcountyhistory.org/rockspringsimage.html

Tharp, T.M., 1987, *Conditions for Crack Propagation by Frost Wedging*, Geological Society of America, Bulletin, v. 99, No. 1, pp. 94-102

Ulrich, E.O., 1916, *Correlation by Displacements of the Strand-Line and the Function and Proper Use of Fossils in Correlation*, Bulletin of the Geological Society of America, pp. 451-490

Photo of the Baraboo Hills, showing the purple quartzite. Photo was taken along Wisconsin Route 33, looking west.

Photo was taken on October 13, 2013 by Steven D.J. Baumann.

Standard printable version

Stratigraphic Column and Ages of the Huronian Supergroup and the Chocolay Group

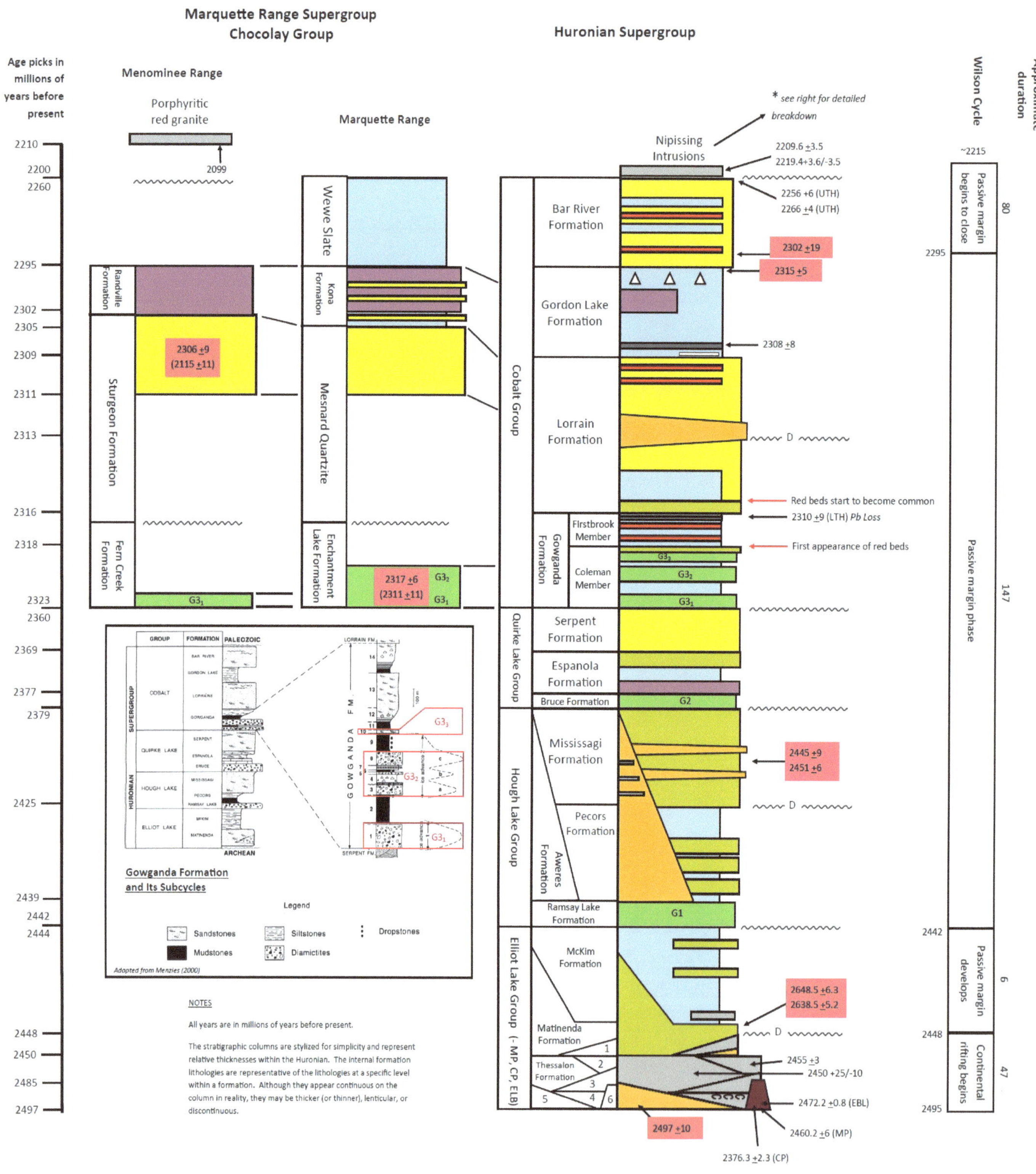

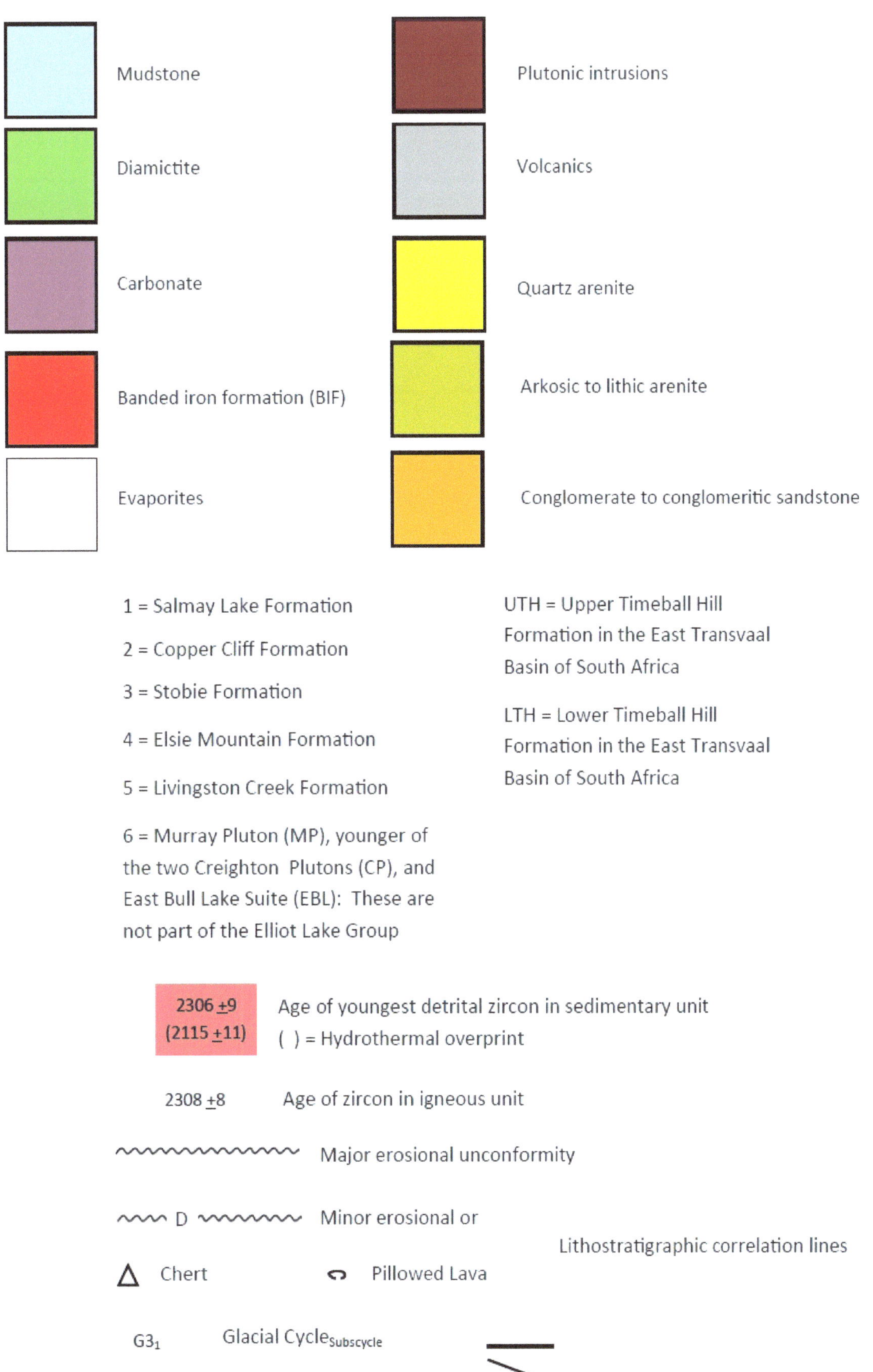

Mudstone
Diamictite
Carbonate
Banded iron formation (BIF)
Evaporites
Plutonic intrusions
Volcanics
Quartz arenite
Arkosic to lithic arenite
Conglomerate to conglomeritic sandstone
1 = Salmay Lake Formation
2 = Copper Cliff Formation
3 = Stobie Formation
4 = Elsie Mountain Formation
5 = Livingston Creek Formation
6 = Murray Pluton (MP), younger of
the two Creighton Plutons (CP), and
East Bull Lake Suite (EBL): These are
not part of the Elliot Lake Group
UTH = Upper Timeball Hill
Formation in the East Transvaal
Basin of South Africa
LTH = Lower Timeball Hill
Formation in the East Transvaal
Basin of South Africa
2306 ±9
(2115 ±11)
Age of youngest detrital zircon in sedimentary unit
( ) = Hydrothermal overprint
2308 ±8
Age of zircon in igneous unit
Major erosional unconformity
D
Minor erosional or
Lithostratigraphic correlation lines
Chert
Pillowed Lava
G3₁
Glacial Cycle Subscycle

* <u>Dates obtained for the Nipissing Intrusions</u>

Actual dates (in Ma)	Minimum possible (in Ma)	Maximum possible (in Ma)	Source
2217$\pm$1.6	2215.4	2218.6	Andrews et al., 1986
2219.4+3.6/-3.5	2215.9	2223.0	Corfu and Andrews, 1986
2217.8+6/-3	2214.8	2223.8	Buchan et al., 1994
2215$\pm$1	2214	2216	Bleeker et al., 2015
2217.2$\pm$4	2213.2	2221.2	Noble and Lightfoot, 1992
2209.6$\pm$3.5	2206.1	2213.1	

Minimum age based off actual dates

Minimum age based off minimum possible

Maximum age based off actual dates

Maximum age based off Maximum possible

Based on the above chart we can tell that the Nipissing Intrusions were not a single event. Ages are outside if margins of error for different samples (see sources). At best (average of 2219.4 and 2209.6 in actual date column) shows we have a depositional spread of 9.8 million years. Based off the averages of our minimum possible (2206.1) and our maximum possible (2223.8), we have a depositional spread of 17.7 million years.

Although we have a handful of dates, we do not have enough dates to firmly state the age range of the Nipissing. If such work is ever done, I predict we will see an early (2225-2222Ma), middle (2219-2216Ma), and a late (2211-2207Ma) set of pulses. A span of about 18 million years.

References

Andrews, A.J., Masliwec, A., Morris, W.A., Owsiacki, L, and York D., 1986, The silver deposits of the Cobalt Group, Ontario Canada: an experiment in age determinations employing radiometric and paleomagnetic measurements. Canadian Journal of Earth Science, 23, 1507-1518

Ayuso, R.A., et al., 2018, New U-Pb zircon ages for rocks from the granite-gneiss terrane in northern Michigan: Evidence for events at ~3750, 2750, and 1850 Ma: Institute on Lake Superior Geology, Proceedings of the 64th annual meeting, Part 1, program and abstracts

Baumann, S.D.J., et al., 2011, Preliminary redefinition of the Cobalt Group (Huronian Supergroup), in the Southern Geologic Province, Ontario, Canada, Midwest Institute of Geosciences and Engineering, G-012011-2A

Buchan, K.L., Mortensen, J.K., and Card, K.D., 1994, Integrated paleomagnetic and U-Pb geochronologic studies of mafic intrusions in the southern Canadian Shield: implications for the Early Proterozoic polar wonder path. Precambrian Research, 69, 1-10

Bleeker, W., Kamo, S, and Ames, D., 2013, New field observations and U-Pb age data for footwall (target) rocks at Sudbury: towards a detailed cross-section through the Sudbury Structure., Large meteorite impacts and planetary evolution V, Geological Society of America, special papers 3112

Bleeker, W., Kamo, S.O., Ames, D.E., and Davis, D., 2015, New field observations and U-Pb ages in the Sudbury area: toward a detailed cross-section through the deformed Sudbury Structure. In "Targeted geoscience initiative 4: Canadian Nickel-Copper Platinum Group elements-Chromium ore-systems-fertility, pathfinders, new and revised models. Geological Survey of Canada, open file 7856, 153-166

Clough, C.E., and Hamilton, M.A., 2017. Matachewan LIP revisited: a revised high resolution U-Pb age for the East Bull Lake Intrusion and associated units. In: "GAC-MAC technical program, T3: Archean cratons and their rifted margins". Secular evolution and metallogeny (Poster) Kingston, Ontario

Craddock, J.P., et al., 2013, Detrital zircon geochronology and provenance of the Paleoproterozoic Huron (~2.4-2.2 Ga) and Animikie (`2.2-1.8 Ga) Basins, Southern Superior Province, Journal of Geology, 121 (2013) 623-644

Corfu, F., and Andrews, A.J., 1986, A U-Pb age for mineralized Nipissing diabase, Gowganda, Ontario. Canadian Journal of Earth Science 23(1) 107-109

Easton, R.M., and Heman, L.M., 2011. Detrital zircon geochronology of Matinenda Formation sandstones (Huronian Supergroup) at Elliot Lake, Ontario: implications for uranium mineralization. Institute on Lake Superior Geology, Proceedings of the 57th ILSG annual meeting, Part 1, 31-32

Hill, C.M., 2019, Sedimentology lithostratigraphy, and geochronology of the Paleoproterozoic Gordon Lake Formation, Huronian Supergroup, Ontario Canada. University of Western Ontario, PhD dissertation

Long, D.G.F., et al., 2011, Laterally extensive modified placer gold deposits in the Paleoproterozoic Mississagi Formation, Clement and Pardo Townships, Ontario, Canadian Journal of Earth Science, 48 (2011) 779-792

Menzies, J., 2000, Microstructures in Diamictites of the Lower Gowganda Formation (Huronian), Near Elliot Lake, Ontario: Evidence for Deforming-Bed Conditions at the Grounding Line?, Journal of Sedimentary Research, 70(1) (2000) 210-216

Noble, S.R., and Lightfoot, P.C., 1992, U-Pb baddeleyite ages of the Kerns and Triangle Mountain intrusions, Nipissing diabase, Ontario. Canadian Journal of Earth Science 29, 1424-1429

Rainbird, R.H., and Davis, W.J., 2006, Detrital zircon geochronology of the western Huronian Basin. Institute on Lake Superior Geology, 52nd aannual meeting, Part I, 55-56

Rasmussen, B., et al., 2013, Correlation of Paleoproterozoic glaciations based on U-Pb zircon ages for tuff beds in the Transvaal and Huronian Supergroups, Earth and Planetary Science Letters, 382 (2013) 173-180

Robertson, J.A., 1976, Geology of the Massey area, Districts of Algoma, Manitoulin, and Sudbury, Ontario Division of Mines, Geosciences Report 136

Rousell, D.H., Brown, G.H. (ed.), 2009, A field guide to the geology of Sudbury Ontario, Ontario Geological Survey, Open File Report 6243

Vallini, D.A., et al., 2006, Age constraints for Paleoproterozoic glaciation in the Lake Superior Region: detrital zircon and hydrothermal xenotime ages for the Chocolay Group, Marquette Range Supergroup

G-092017-1A: Temporal Distribution Chart of the Superiorian and Rhyacian Deposits in Ontario, Michigan, and South Africa

Compiled by: Steven D.J. Baumann (September 2017)

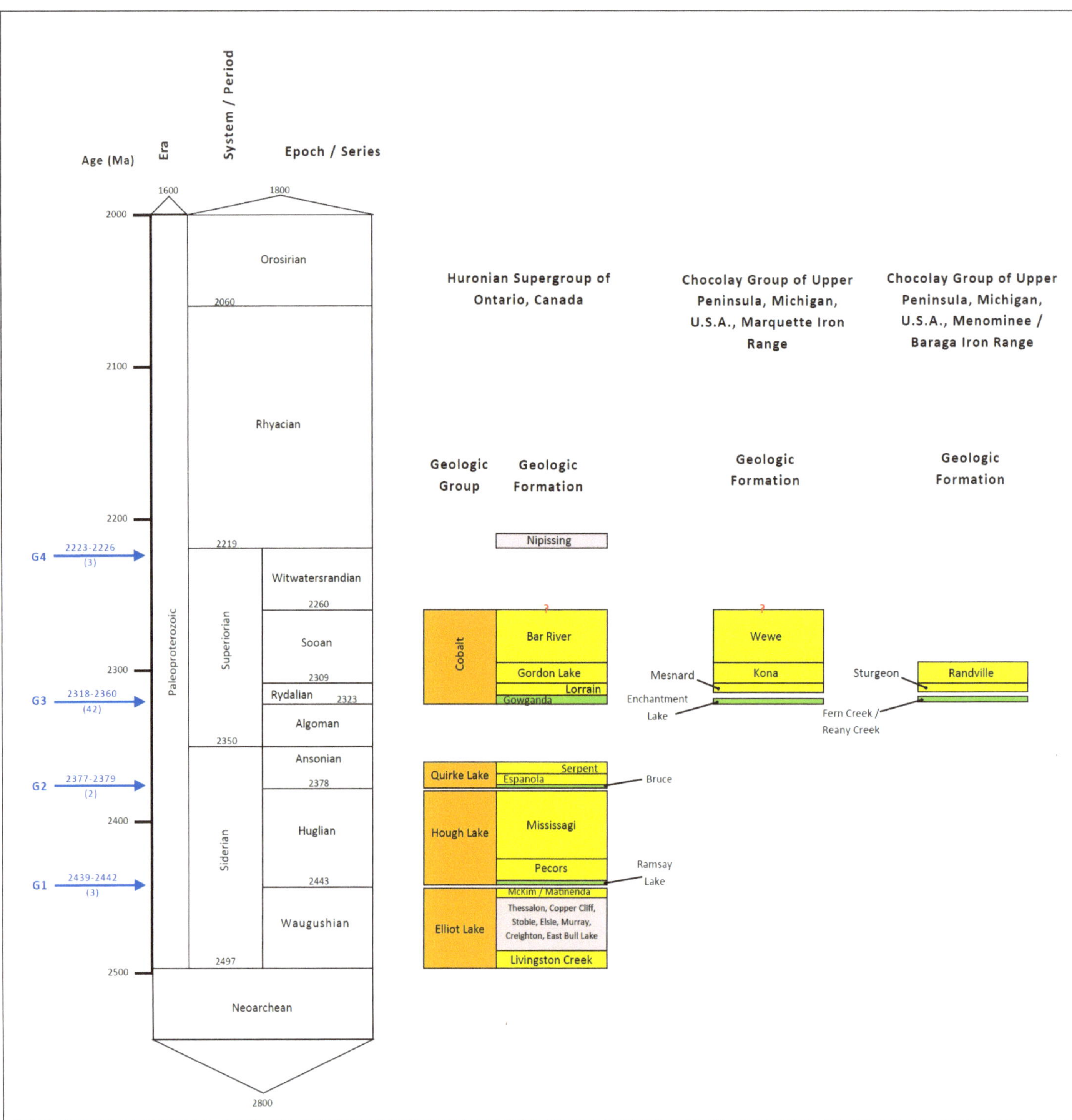

The unfamiliar names that appear for some of the periods and epochs above, are in no way formal. At the time, they were a "work in progress". To see the official suggested divisions, go to the final publication in this book **G-032021-1A:** *Chronology and redefinition of the Mesoproterozoic, Paleoproterozoic, Archean, and Hadean.*

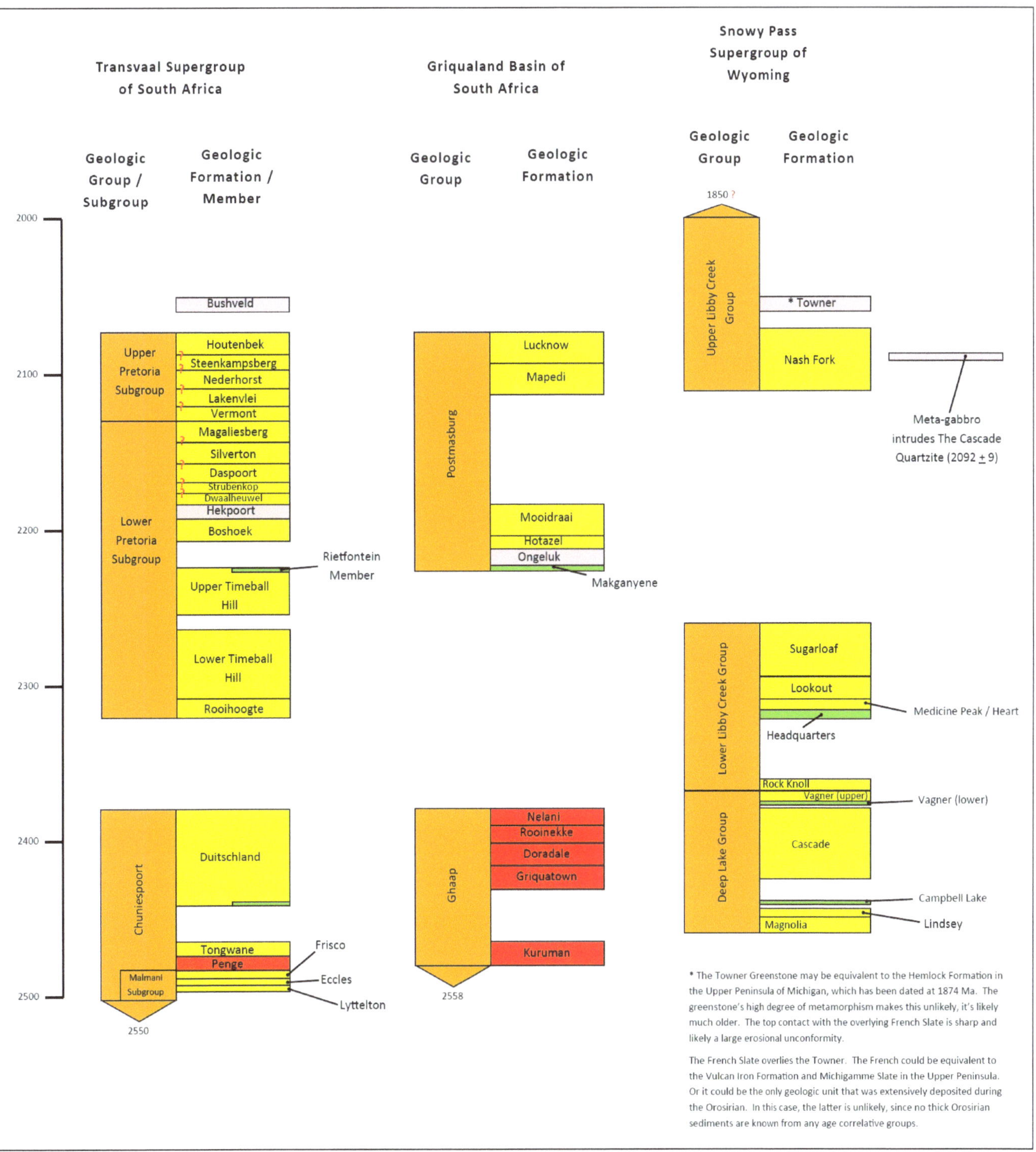

* The Towner Greenstone may be equivalent to the Hemlock Formation in the Upper Peninsula of Michigan, which has been dated at 1874 Ma. The greenstone's high degree of metamorphism makes this unlikely, it's likely much older. The top contact with the overlying French Slate is sharp and likely a large erosional unconformity.

The French Slate overlies the Towner. The French could be equivalent to the Vulcan Iron Formation and Michigamme Slate in the Upper Peninsula. Or it could be the only geologic unit that was extensively deposited during the Orosirian. In this case, the latter is unlikely, since no thick Orosirian sediments are known from any age correlative groups.

Notes

All ages (written as numbers) are in millions of years before present (Ma).

Triangular arrows indicate the geologic unit or timespan continues to the age at the top (or bottom) of the triangular arrow.

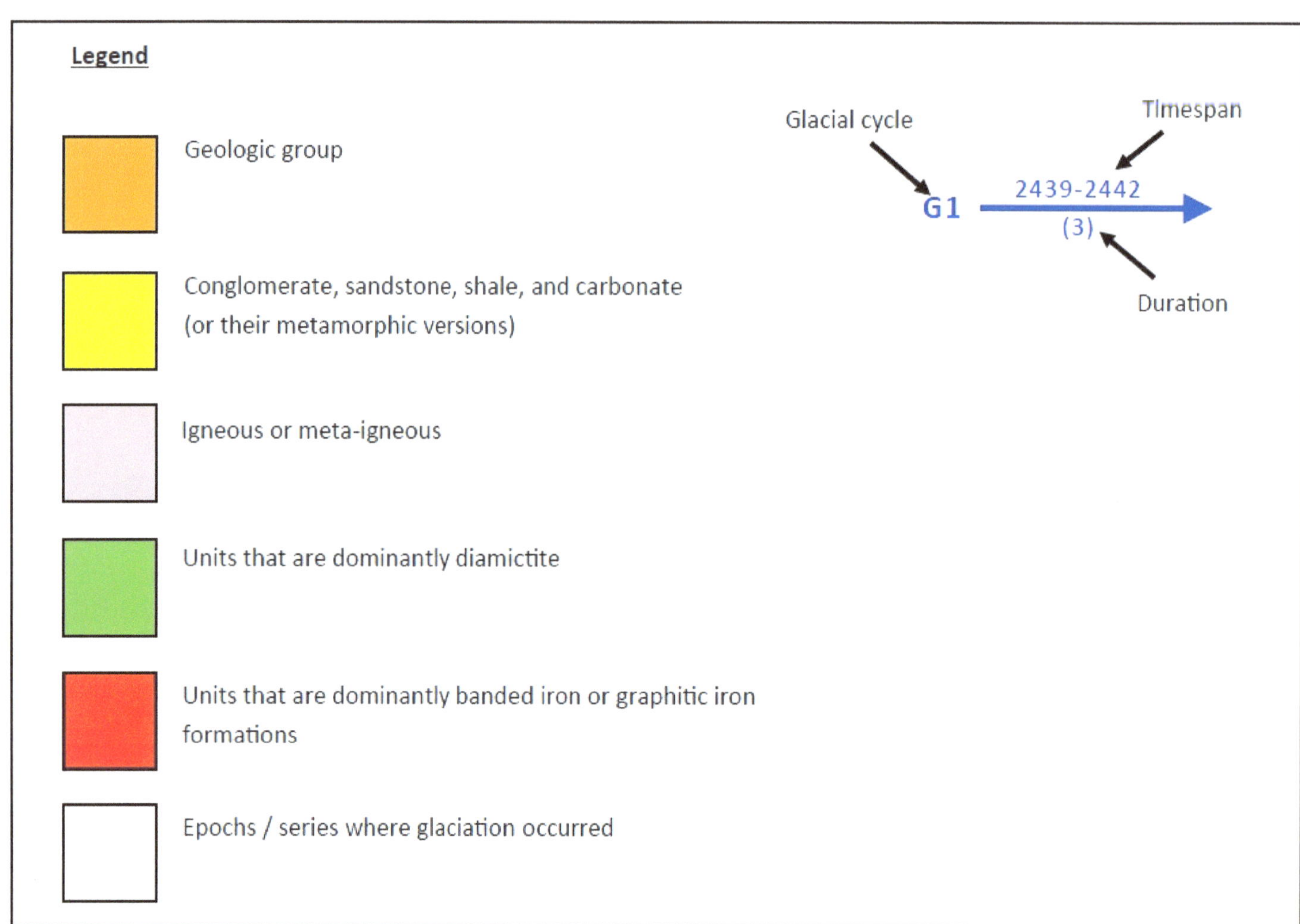

This legend applies to the two previous pages

If this correlation chart is accurate, then the following can be surmised from it.

1) G1 was an extensive glacial cycle. Deposits are found in North America and South Africa. Although extensive, there's no evidence that it was a global glacial event.

2) G2 was not as extensive as G1, it was also shorter. South Africa contains no G2 deposits. It is possible that they were eroded or not deposited at all.

3) G3 has left thick deposits with subcycles in North America, but none are present in South Africa, nor were they deposited. Other non-glacial deposits were being deposited in South Africa while glacial sediments were being deposited in North America. So it is very unlikely that a "snowball earth event" occurred at this time, even though G3 spans 42 million years.

4) G4 is preserved in South Africa but there are none in North America, although they could have been deposited (and eroded), but this seems unlikely. A snowball earth scenario is also not likely here, because deposits after G4 are non-glacial and continuous in South Africa.

5) The Algoman period contains no sediments and appears to represent a global unconformity. There are several possibilities as to why, or a combination of factors. An erosional unconformity formed due to a drop in sea level caused by ice caps or tectonics, or a general uplift of continental crust caused by tectonics. Another mechanism could be a drop in sea level caused by the establishment of new deep ocean trenches and a general lack of younger more buoyant ocean crust. During the Algoman, the breakup of the paleo-continent(s) Kenorland, was well underway. It's possible Kenorland was three smaller, yet close continents.

6) Just before the breakup of Kenorland and the onset of G1, both North America and South Africa appear to have been located at mid-latitudes ($30°$ to $60°$ latitude).

7) Post Kenorland breakup continental positions are not known. Using the above stratigraphy, it looks like North America was slightly closer to a pole than South Africa was during G1. This persisted through G2 and G3, with North America (Superior Craton) likely within $30°$ of a pole. By G4, North America was either closer to the equator than South Africa was, or the local tectonic regime favored erosion in North America and continued deposition in South Africa. Either way, during G4 North America was likely at mid-latitudes.

8) At no time during the Superiorian or the Siderian does there appear to be a snowball earth scenario. Extensive glaciation down to mid-latitudes may have occurred (although rare), a frozen over Earth is not likely during the Paleoproterozoic.

9) Paleomagnetic data from this timespan is practically non-existent. In order to form a better picture, we need that data.

References

2012, GSA Geologic Time Scale, Geological Society of America, v. 4.0

2017, International Chronostratigraphic Chart, International Commission on Stratigraphy, v 2017/02

Baumann, S.D.J., et al., 2011, Preliminary redefinition of the Cobalt Group (Huronian Supergroup), in the Southern Geologic Province, Ontario, Canada, Midwest Institute of Geosciences and Engineering, G-012011-2A

Baumann, S.D.J., 2017, Stratigraphic columns and ages of the Huronian Supergroup and Chocolay Group, Midwest Institute of Geosciences and Engineering, G-082017-1A

Baumann, S.D.J., 2017, Redefining the Paleoproterozoic Siderian and Rhyacian Periods, Midwest Institute of Geosciences and Engineering, G-082017-2A

Bekker, A., 2001, Chemostratigraphy of the early Paleoproterozoic carbonate successions (Kaapvaal and Wyoming cratons), Virginia Polytechnic Institute and State University, doctorate dissertation

Bekker, A., 2003, Chemostratigraphy of Paleoproterozoic carbonate successions of the Wyoming Craton: tectonic forcing of biogeochemical change?, Journal of Precambrian Research, 120 (2003) 279-325

Catuneanu, O. and Eriksson, P.G., 2002, Sequence Stratigraphy of the Precambrian Rooihoogte-Timeball Hill rift succession, Transvaal Basin, South Africa, Journal of Sedimentary Geology, 147 (2002) 71-88

Craddock, J.P., et al., 2013, Detrital zircon geochronology and provenance of the Paleoproterozoic Huron (~2.4-2.2 Ga) and Animikie (`2.2-1.8 Ga) Basins, Southern Superior Province, Journal of Geology, 121 (2013) 623-644

Eriksson, P.G. and Catuneanu, O., Houston, R.S., 1978, A regional study of the rocks of Precambrian age in that part of the Medicine Bow Mountains lying in southeastern Wyoming—with a chapter on the relationship between the Precambrian and Laramide Structure, Geological Survey of Wyoming, Memoir No. 1

Long, D.G.F., et al., 2011, Laterally extensive modified placer gold deposits in the Paleoproterozoic Mississagi Formation, Clement and Pardo Townships, Ontario, Canadian Journal of Earth Science, 48 (2011) 779-792

Lewy, Z., 2009, Early Precambrian banded iron formations: biochemical precipitates from highly evaporated hydrothermal solutions of polar region lakes, Journal of Carbonates and Evaporites, 24 (2009) 1-15

Menzies, J., 2000, Microstructures in Diamictites of the Lower Gowganda Formation (Huronian), Near Elliot Lake, Ontario: Evidence for Deforming-Bed Conditions at the Grounding Line?, Journal of Sedimentary Research, 70(1) (2000) 210-216

Rasmussen, B., et al., 2013, Correlation of Paleoproterozoic glaciations based on U-Pb zircon ages for tuff beds in the Transvaal and Huronian Supergroups, Earth and Planetary Science Letters, 382 (2013) 173-180

Robertson, J.A., 1976, Geology of the Massey area, Districts of Algoma, Manitoulin, and Sudbury, Ontario Division of Mines, Geosciences Report 136

Rousell, D.H., Brown, G.H. (ed.), 2009, A field guide to the geology of Sudbury Ontario, Ontario Geological Survey, Open File Report 6243

Vallini, D.A., et al., 2006, Age constraints for Paleoproterozoic glaciation in the Lake Superior Region: detrital zircon and hydrothermal xenotime ages for the Chocolay Group, Marquette Range Supergroup

QAP plots of granitic rocks of northeastern Wisconsin

G-112019-1A **Steven D.J. Baumann**

Midwest Institute of Geosciences and Engineering

10 Years (2009-2019)

Photo on the previous page is of the Hoskin Lake Granite sample collected by
Steven D.J. Baumann from US-141 on June 19, 2019. Photo was taken by
Steven D.J. Baumann, during lab analysis on July 2, 2019.

**First edition
November
2019**

Abstract

30 samples were taken from the Wausau-Pembine Terrane of the Penokean Orogeny of all the major phaneritic plutons, except for the Bush Lake Granite. The following charts were constructed to better constrain the lithology of each pluton. More work still needs to be done, but it is becoming apparent that the Athelstane Quartz Monzonite may actually be two separate intrusions derived from the same magma source. The Baxter Hollow Granite, is not considered to be Penokean herein because of its post-assemblage date. It could still be considered a late stage pluton that was injected as the subducting plate was completely consumed back into the mantle.

Base QAP Plot for all samples

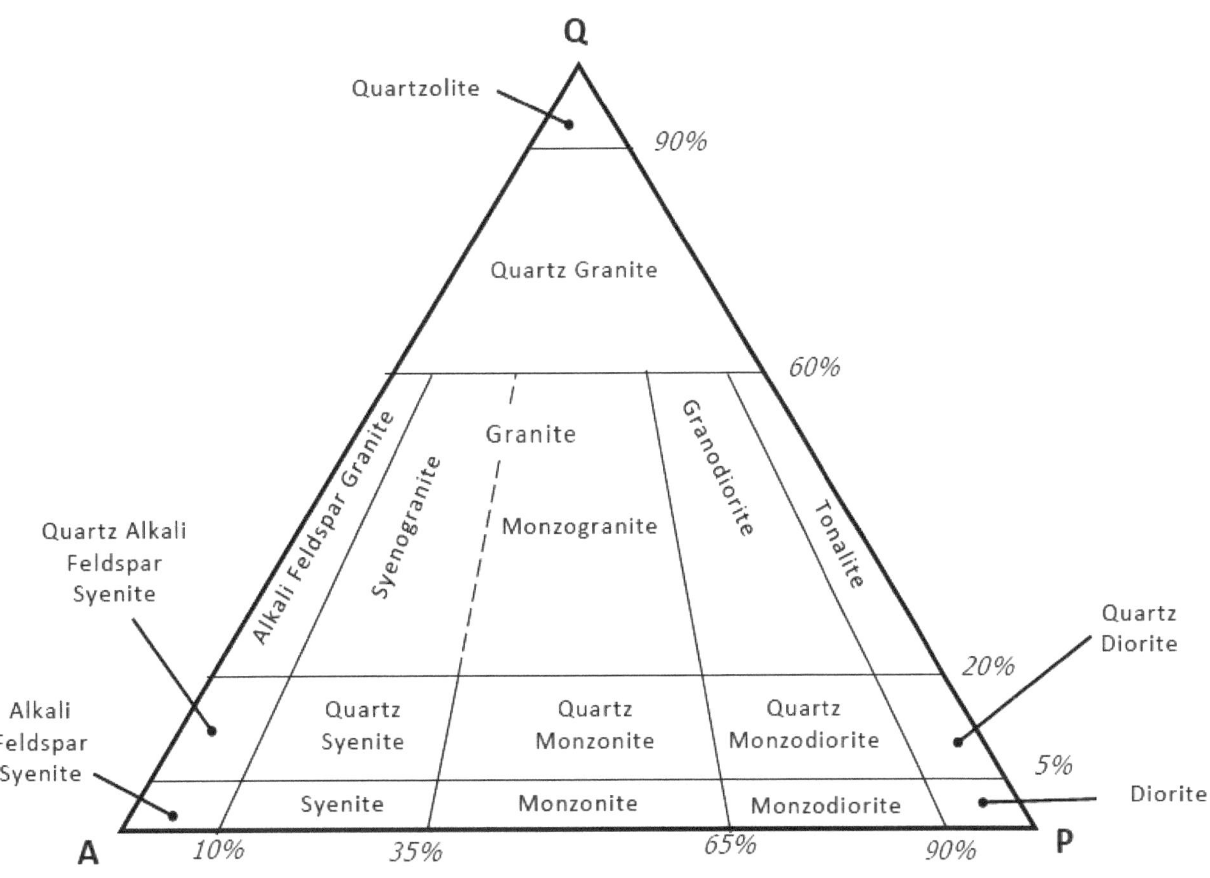

Adapted from Streckeisen, 1974

Amberg Granite (Not Penokean and not in the Pembine-Wausau Terrane)

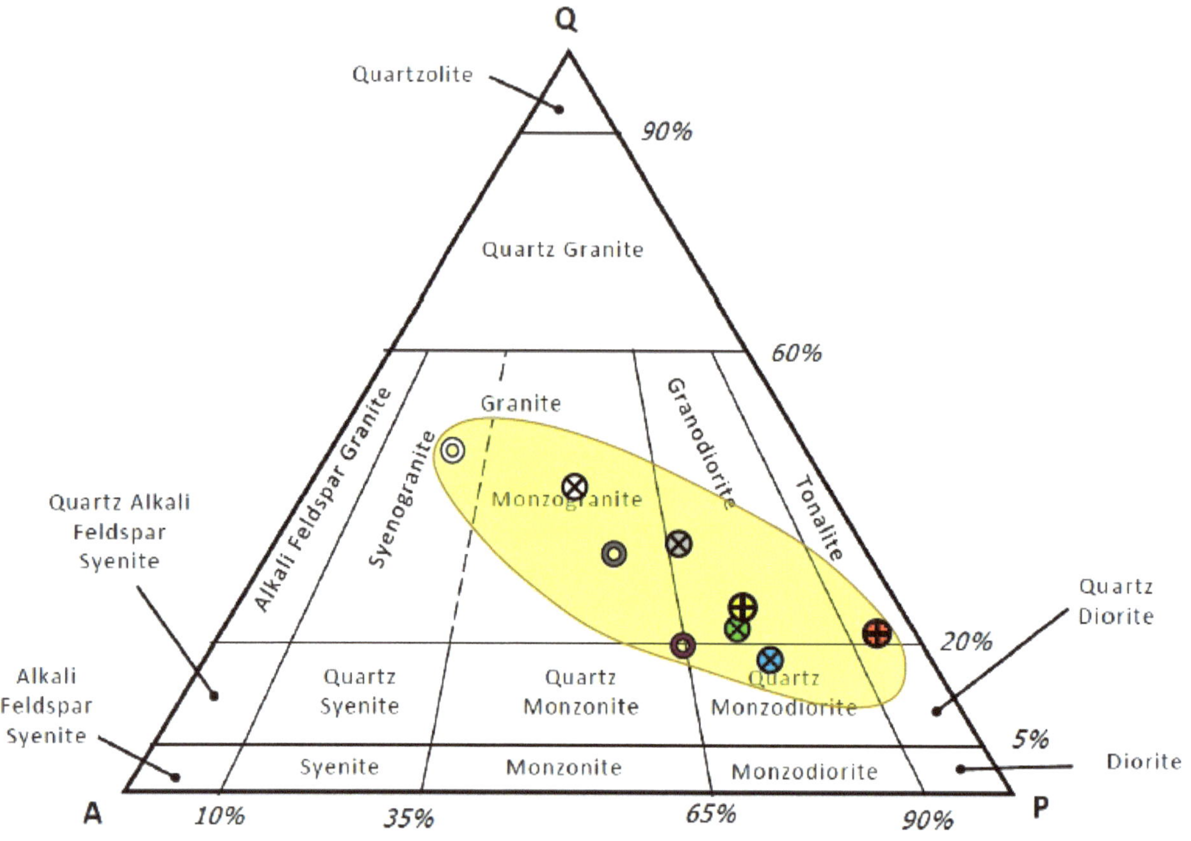

The above QAP ternary plot includes the W12B sample of Sims, et al. (1993) but the extent is based off of my samples and sample W12B.

My sample: 063019-4

USGS sample W12B

My sample: 071919-1

My sample: 071919-6

My sample: 060819-2

My sample: 081019-4 (dike)

My sample: 081019-6 (dike)

My sample: 081019-7 (dike)

My sample: 081019-9 (dike)

Age:

1759$\pm$19Ma
(Peterman et al., 1985)

1754$\pm$11Ma
(Holm et al., 2005)

Athelstane Quartz Monzonite

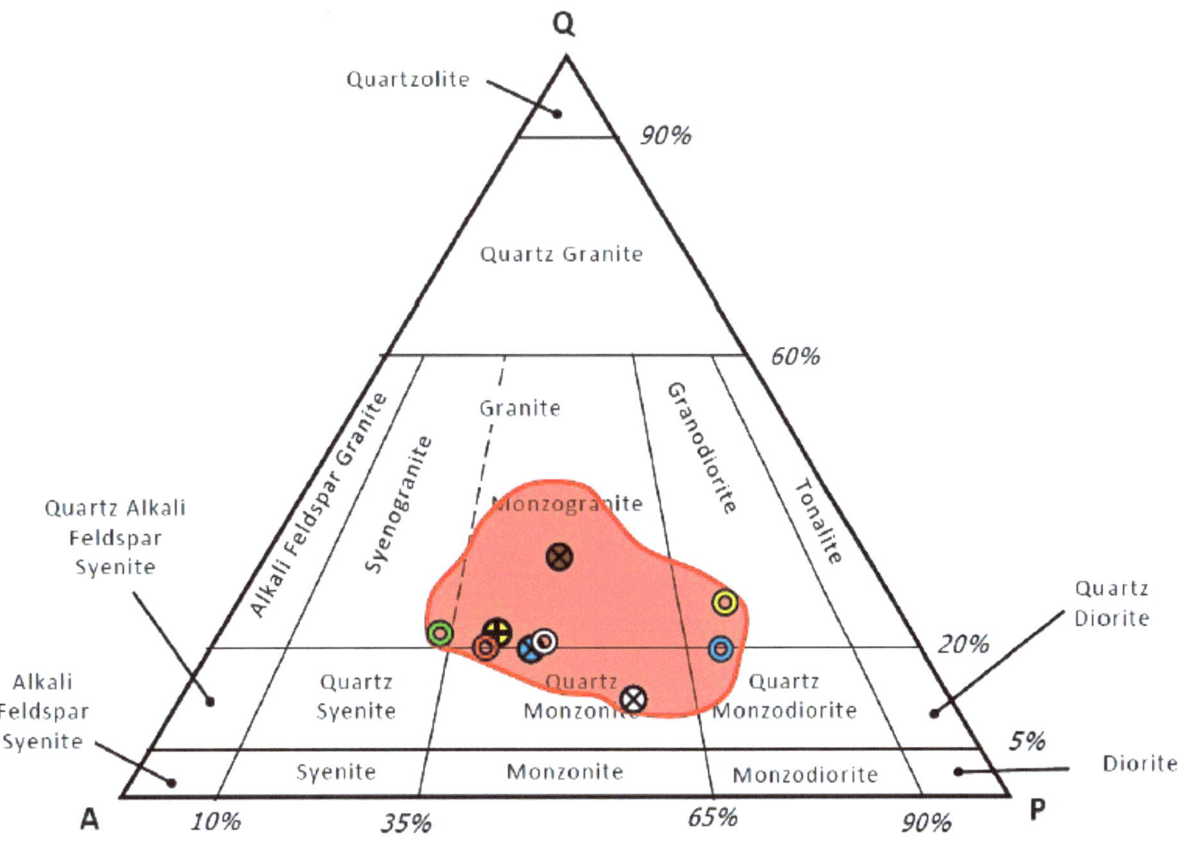

The above QAP ternary plot is adapted from the samples of Sims, et al. (1993) and my own samples.

My sample: 062919-6 (Felsic dike)

My sample: 063019-5

My sample: 071919-2

My sample: 071919-4

My sample: 071919-5

My sample: 081019-1

My sample: 081019-2

My sample: 081019-3

My sample: 081019-8

Age:

1836$\pm$15Ma

(Simms, 1990)

<u>Baxter Hollow Granite</u>   (Not Penokean)

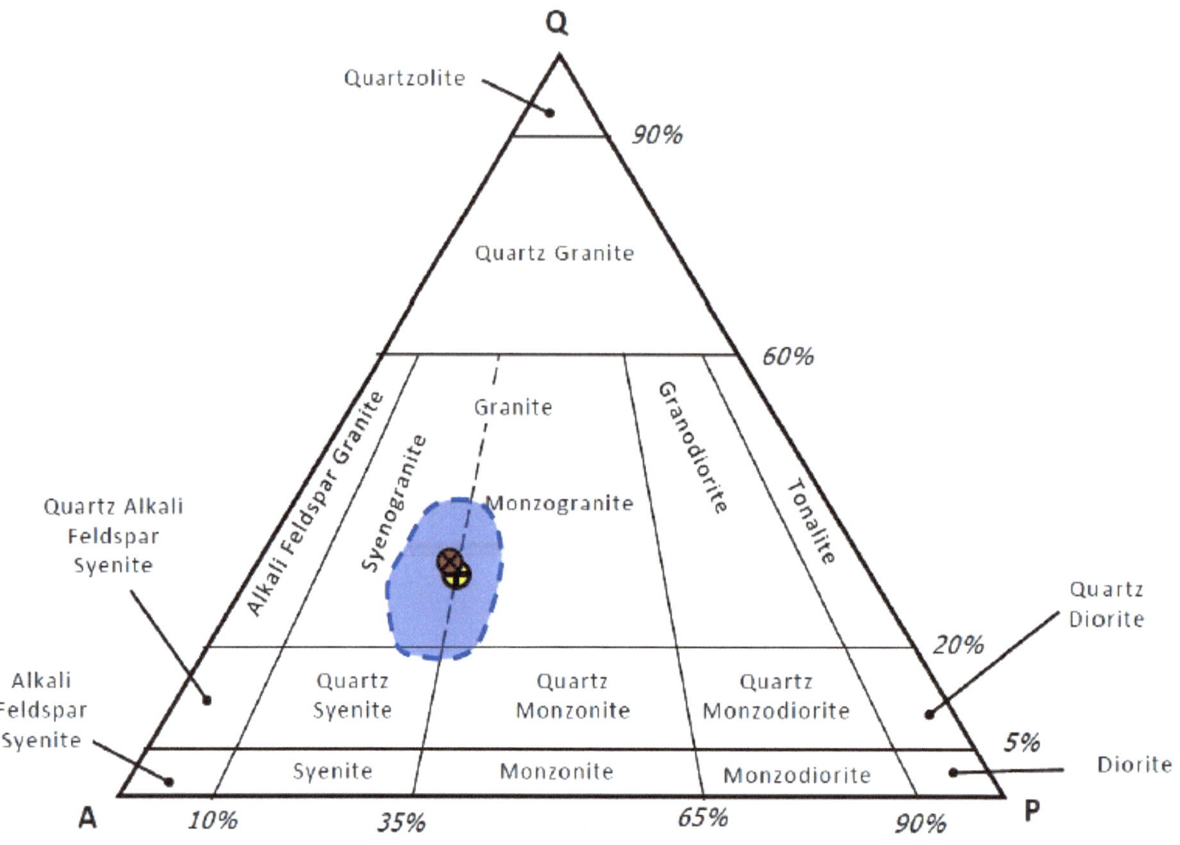

*   *The extent of the circle is a slight expansion of the low and high values of Gates (1942) three main sample types. It could be slightly larger or smaller.*

⊕   My sample: 081719-1

⊗   Mean average (Gates, 1942)

Age:

1749±12Ma
(Van Wyck, 1995)

1746±3Ma
(Van Schmus et al., 2001)

<u>Bush Lake Granite</u>

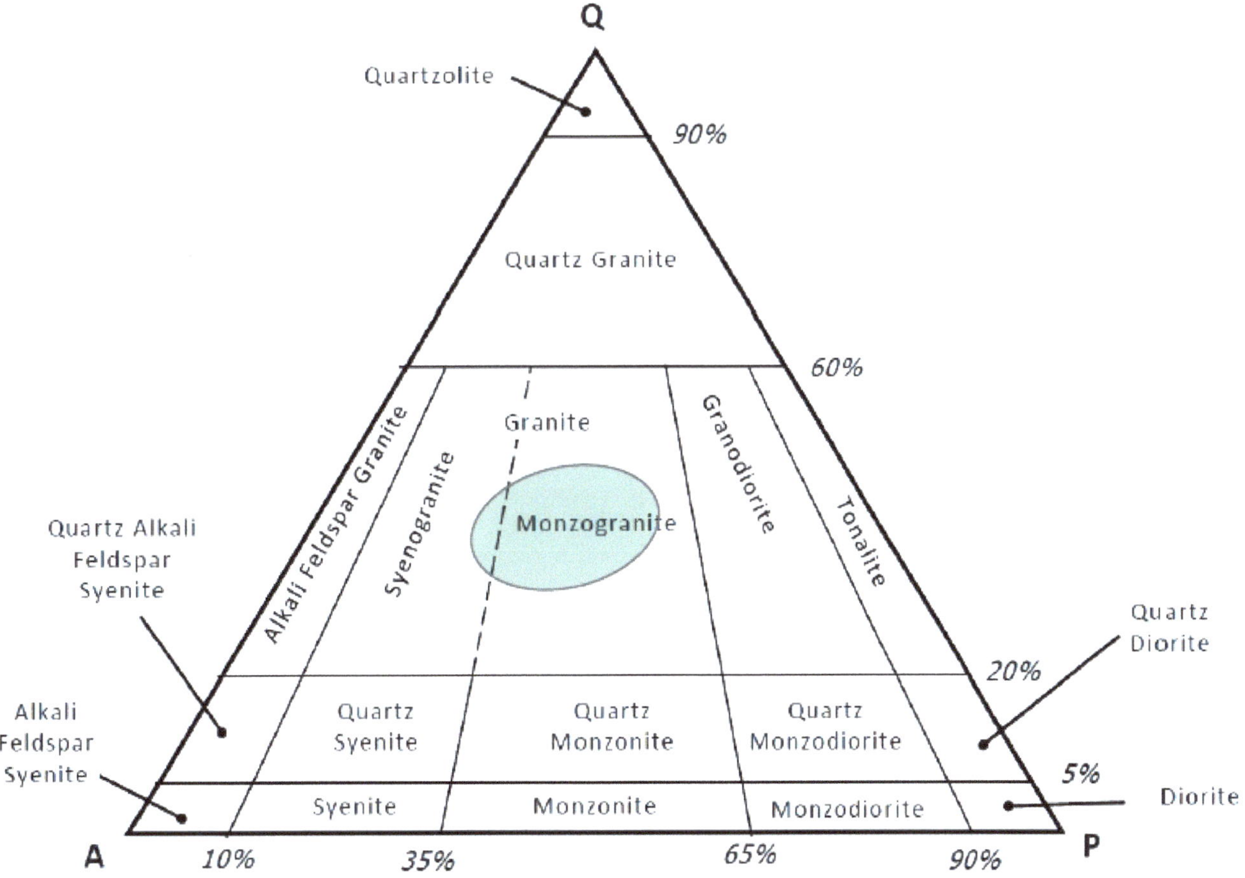

*   *The Bush Lake Granite was not visited or sampled by me. The above QAP ternary plot is adapted from the samples of Sims et al. (1992) and Sims, et al. (1993).*

Age:

Not directly dated

<u>Dunbar Gneiss</u>

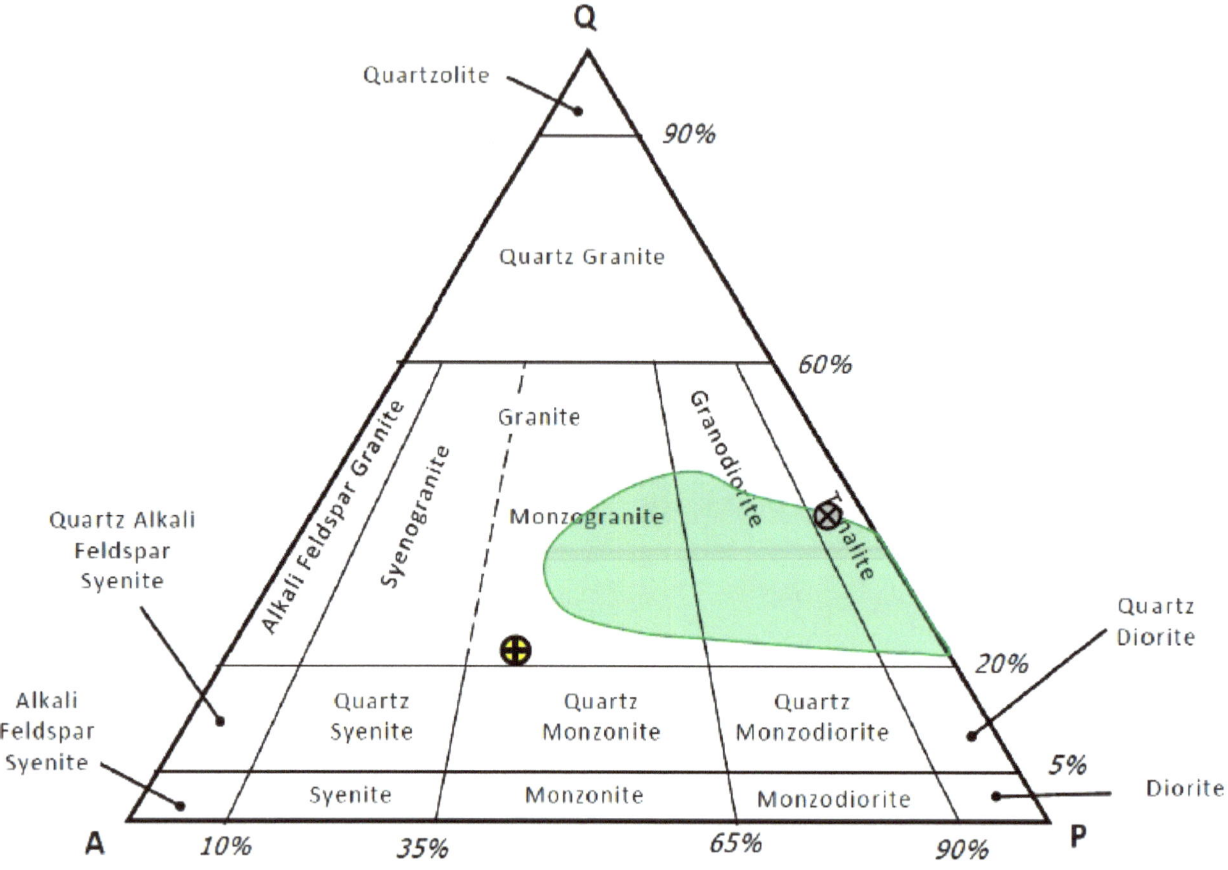

The above QAP ternary plot is adapted from the samples of Sims, et al. (1992). As you can see, my sample of the felsic dike plotted way outside of the limits and it could likely be a dike of the Newingham Tonalite, or less likely the Athelstane Quartz Monzonite.

⊗   My sample: 062919-6 (Main part)

⊕   My sample: 062919-6 (Felsic dike part)

Age:

1862$\pm$5Ma
(Sims, 1990)

1862$\pm$4Ma
(Peterman et al., 1985)

Hoskin Lake Granite

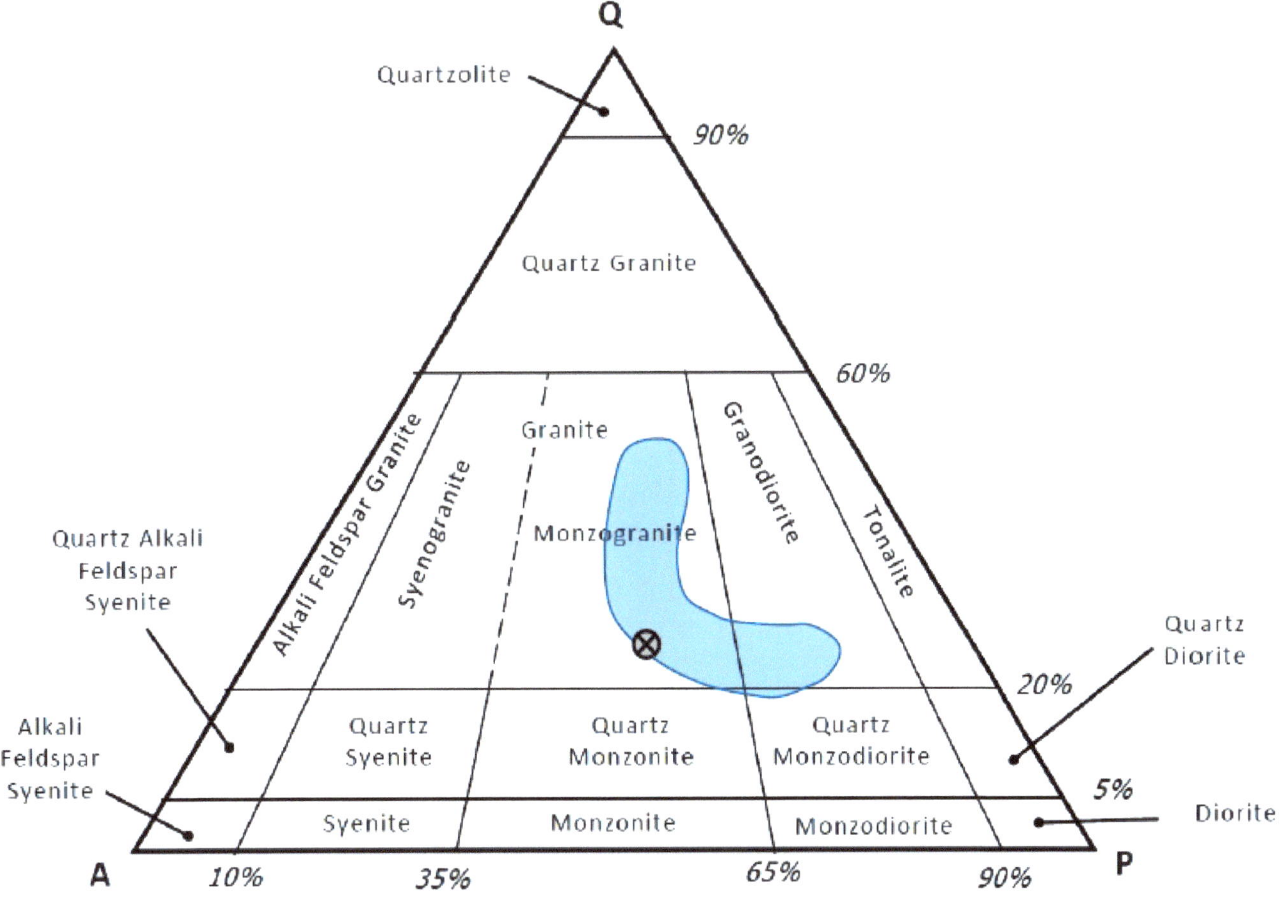

The above QAP ternary plot is adapted from the samples of Sims, et al. (1992) as well as my own sample.

⊗ My sample: 062919-4

Age:

1861±32Ma
(Sims, 1990)

Marinette Quartz Diorite

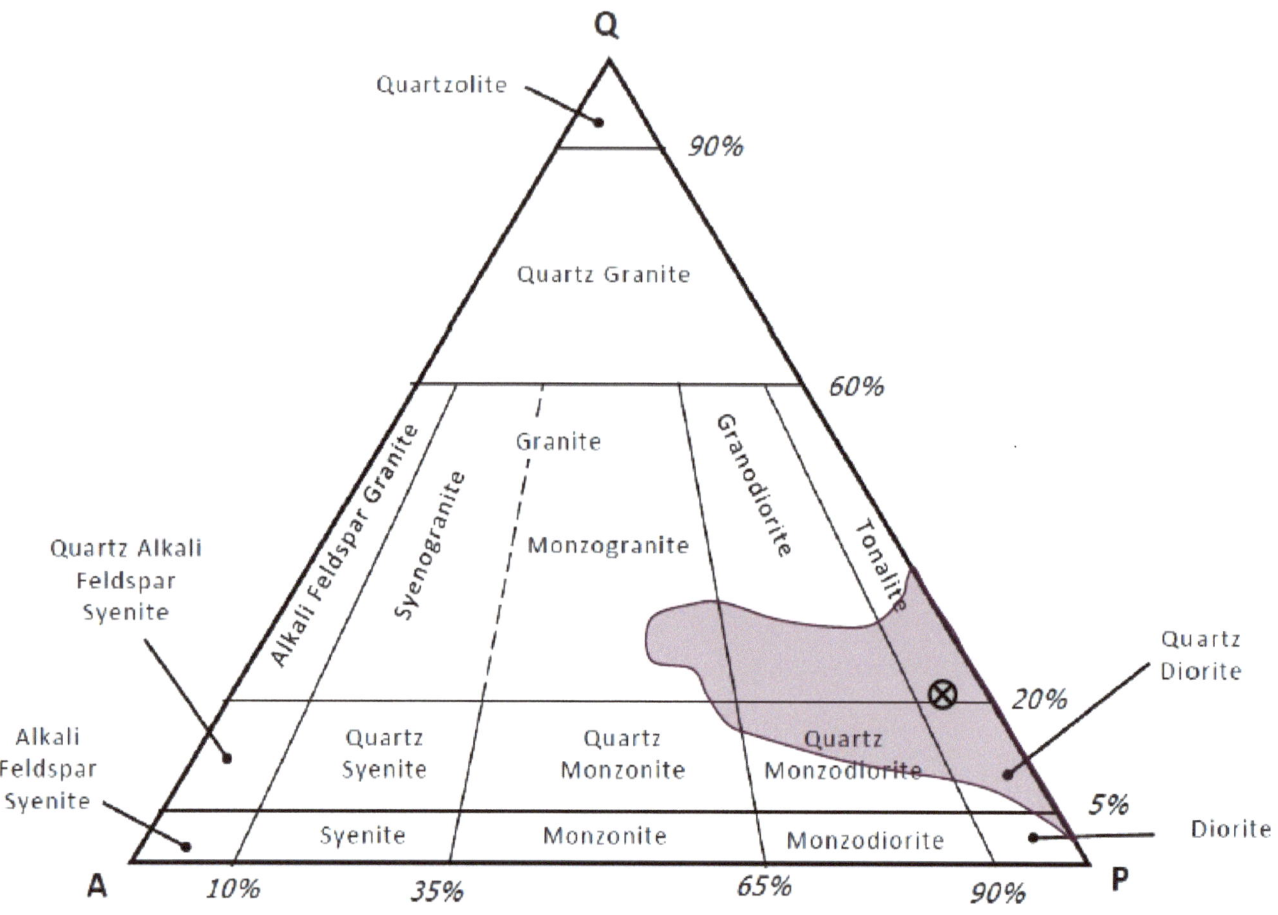

The above QAP ternary plot is adapted from the samples of Sims, et al. (1992) as well as my own sample.

⊗ My sample: 062919-7

Age:

1857±6Ma
(Sims and Schulz, 1993)

1862±15Ma
(Sims et al., 1992)

Newingham Tonalite

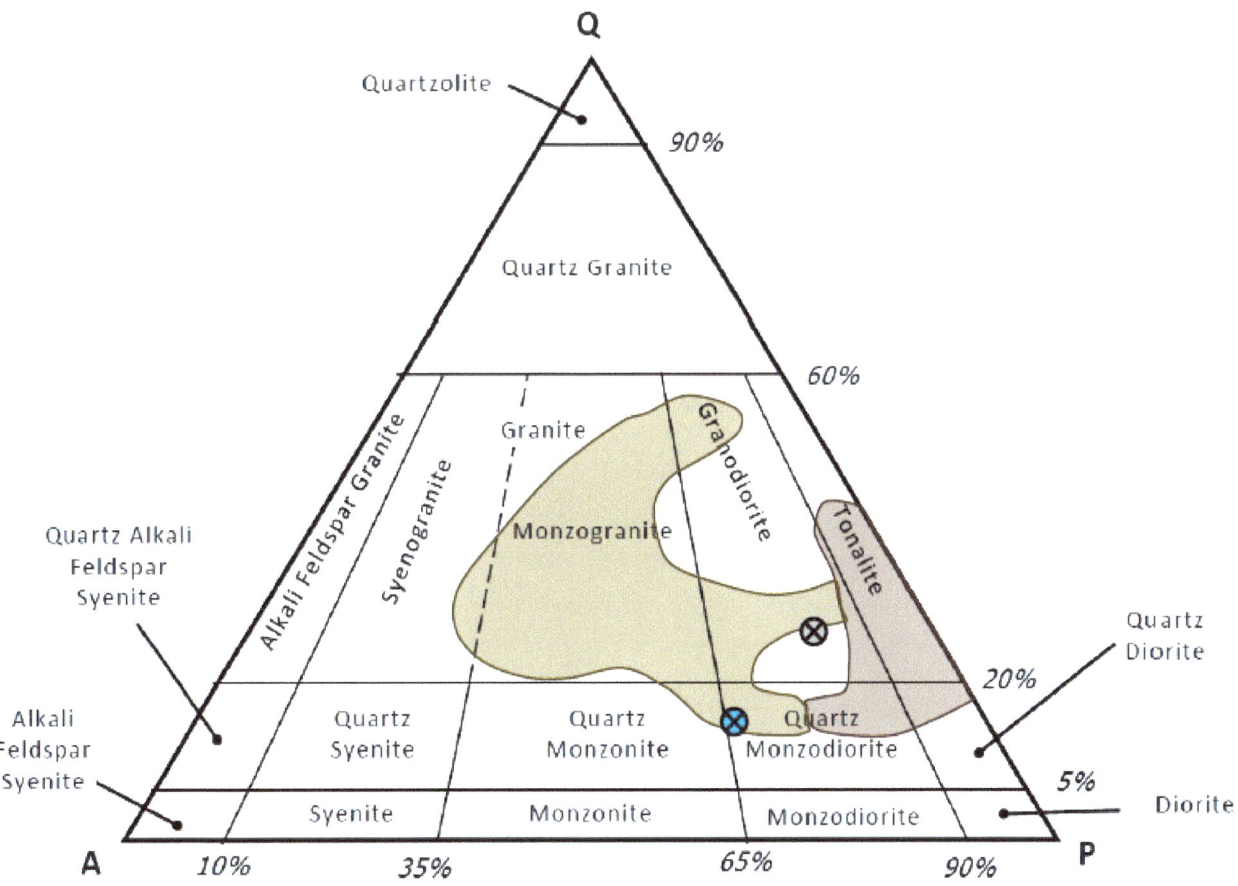

The above QAP ternary plot is adapted from the samples of Sims, et al. (1992) as well as my own samples.

⊗ My sample: 061019-4   ⊗ My sample: 063019-2

Megacryst facies   Tonalite facies

Age:

1861±40Ma
(Sims, 1990)

Spikehorn Creek Granite

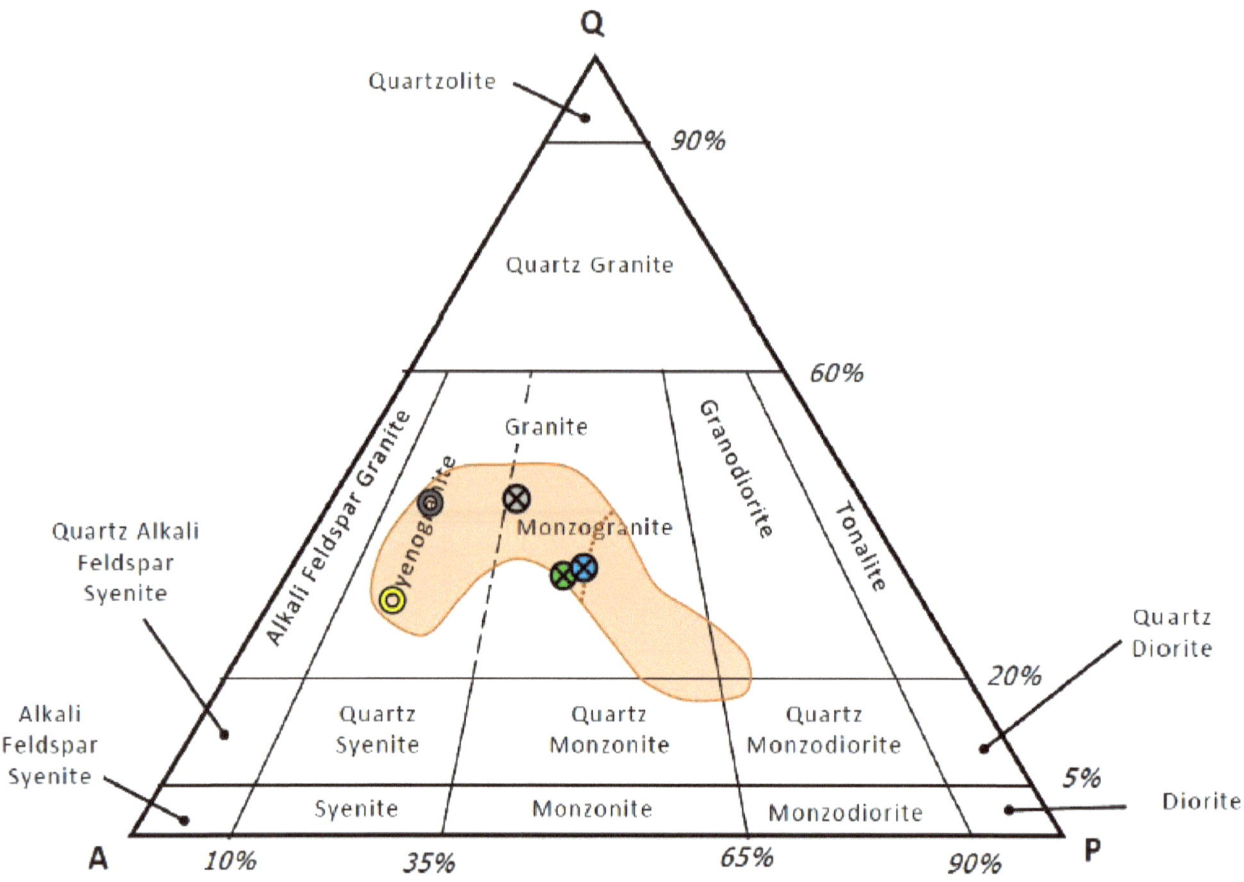

The above QAP ternary plot is adapted from the samples of Sims, et al. (1992), as well as my own samples.

⊗ My sample: 060819-7   ⊗ My sample: 060819-9

⊗ My sample: 060819-11

◉ My sample: 061019-2   ◉ My sample: 061019-3

⋯ Dividing line between old calculated (to the right) and new calculated (to the left) extent

Age:

1836±6Ma
(Peterman et al., 1985)

1835±5Ma
(Sims, 1990)

Twelve Foot Falls Quartz Diorite

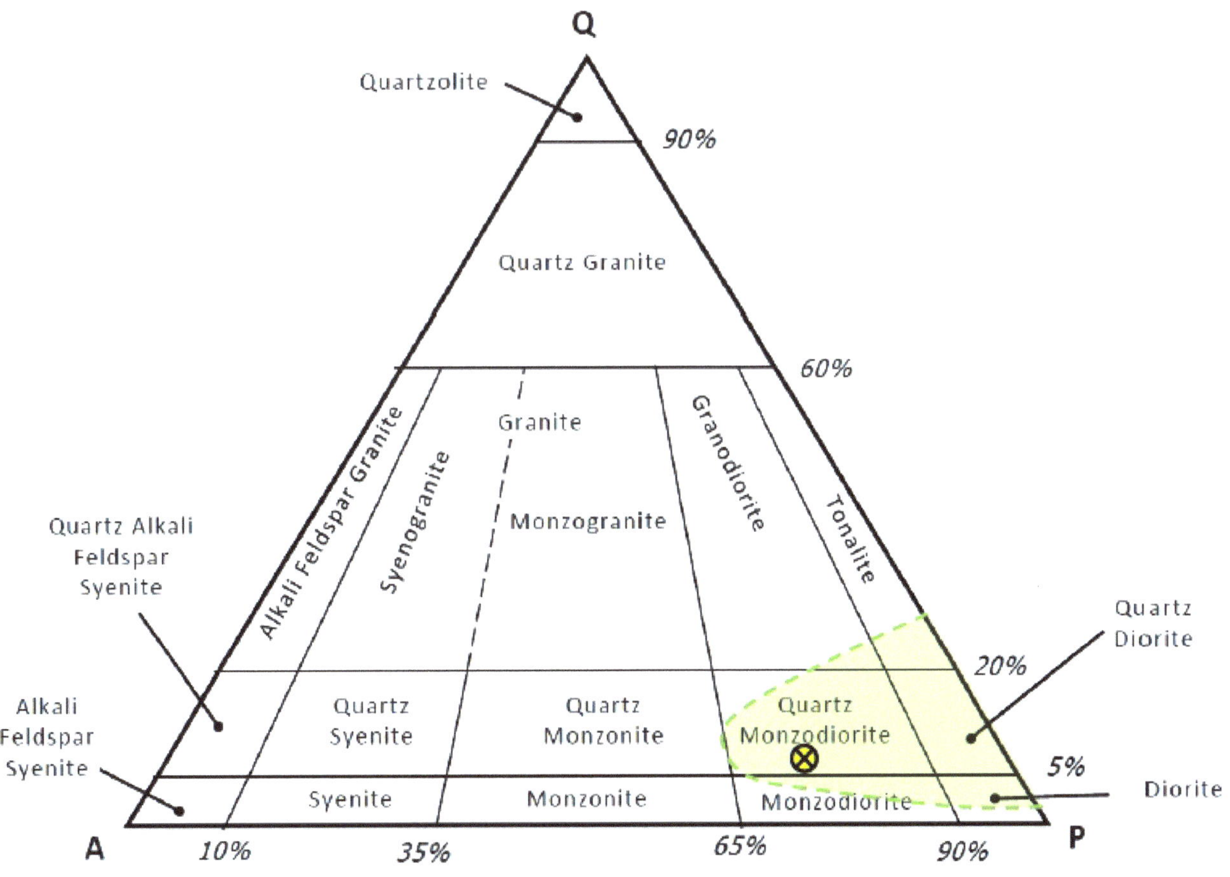

 My sample: 062919-8

Likely extent of Twelve Foot Falls Quartz Diorite

Age:

Not directly dated

Average of all Penokean Plutons in Northeast Wisconsin

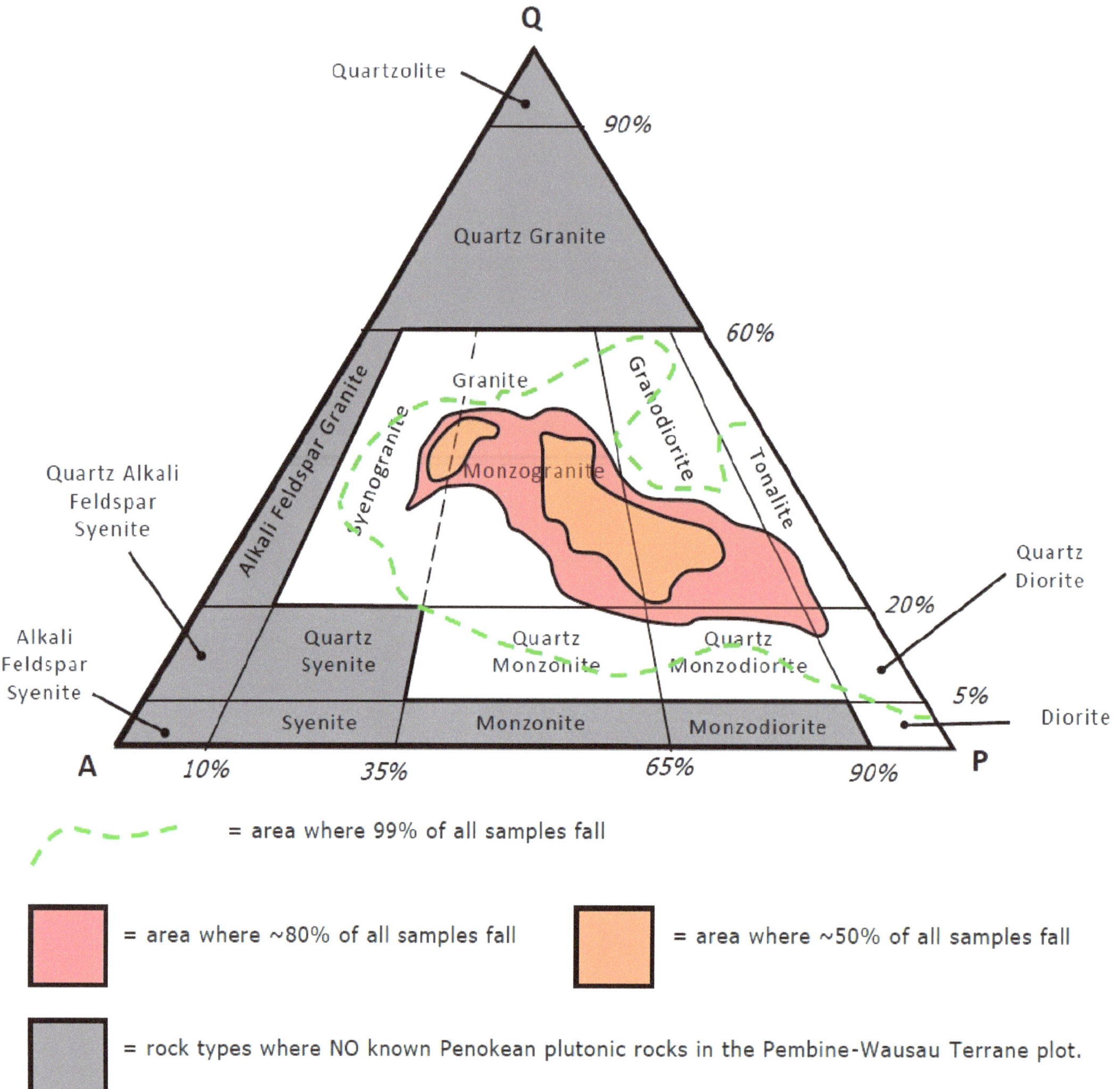

= area where 99% of all samples fall

= area where ~80% of all samples fall

= area where ~50% of all samples fall

= rock types where NO known Penokean plutonic rocks in the Pembine-Wausau Terrane plot.

* *The above chart does not include the Amberg Granite nor does it include the Baxter Hollow Granite, since neither is Penokean. The Amberg has been dated at 1759±19Ma (Peterman, et al., 1985) and 1754±11Ma (Holm, et al., 2005). The Baxter Hollow Granite (about 180 miles to the south-southwest and outside the Pembine-Wausau Terrane) has been dated to 1749±12Ma (Van Wyck, 1995). This is much younger than the youngest dated Penokean pluton in the area, the Spikehorn Creek / Bush Lake at 1835±5Ma (Sims, 1990).*

* *The above chart is a combination of the previous QAP ternary plots based off my samples and the samples of Sims, et al. (1993).*

References:

Gates, R.M., 1942. Baxter Hollow Granite cupola. American mineralogist, No. 27, Vol. 10, p. 699-711

Holm, D.K., Van Schmus, W.R., MacNeill, L.C., Boerboom, T.J., Schweitzer, D., and Schneider, D., 2005. U-Pb zircon geochronology of the Paleoproterozoic plutons from the northern midcontinent, USA: Evidence for subduction flip and continued convergence after Geon 18 Penokean Orogenesis. Geological Society of America, Bulletin 2005, No. 117, Vol. 3-4, p. 259-279

Peterman, Z.E., Sims, P.K., Zartman, and R.E., Schulz, K.J., 1985. Middle Proterozoic uplift events in the Dunbar Dome of northeastern Wisconsin USA.

Schulz, K.J., and Bjornerund, M., 2018. Field Trip 4, Granitoid Rocks of the Pembine-Wausau Terrane in Northeast Wisconsin. Institute on Lake Superior Geology 64th Annual Meeting, Part 2, p. 106-124

Sims, P.K., 1990. Geologic map of Iron Mountain and Escanaba l°x2°quadrangles, northeastern Wisconsin and northwestern Michigan. U.S. Geological Survey Miscellaneous Investigations Series Map, I-2056

Sims, P.K., 1992. Geologic map of Precambrian rocks, southern Lake Superior region, Wisconsin and northern Michigan. U.S. Geological Survey Map I-2185

Sims, P.K., Schulz, K.J., and Peterman, Z.E., 1992. Geology and geochemistry of the Early Proterozoic rocks in the Dunbar area, northeastern Wisconsin. United Stated Geological Survey, Professional Paper 1517

Sims, P.K., and Schulz, K.J., 1992. Geologic map of Precambrian rocks southern Lake Superior region, Wisconsin and northern Michigan, United States Geological Survey Miscellaneous Investigations Series Map, I-2185

Sims, P.K., and Schulz, K.J., 1993. Geologic map of Precambrian rocks of parts of Iron Mountain and Escanaba 30'x60' quadrangles, northeastern Wisconsin and adjacent Michigan. United States. Geological Survey Miscellaneous Investigations Series Map, I-2356

Sims, P.K., Schulz, K.J., DeWitt, E.D., and Brasaemle, B., 1993. Petrography and Geochemistry of Early Proterozoic Granitoid Rocks in Wisconsin Magmatic Terranes of Penokean Orogen, Northern Wisconsin. United States Geological Survey, Bulletin 1904-J

Streckeisen, A. 1974. Classification and nomenclature of plutonic rocks: Recommendations of the IUGS subcommission on the systematics of igneous rocks. Geologische Rundschau, vol. 63, pp. 773-786.

Van Schmus, W. R.; MacNeill, L. C.; Holm, D. K.; and Boerboom, T. J. 2001. New U-Pb ages from Minnesota, Michigan, and Wisconsin: implications for Late Paleoproterozoic crustal stabilization. Institute on Lake Superior Geology Proceedings, 47th annual meeting (Madison, Wis.), 47, (abstract).

Van Wyck, N., 1995. Major and trace element, common Pb, Sm-Nd, and zircon geochronology constraints on petrogenesis and tectonic setting of pre- and Early Proterozoic rocks in Wisconsin, Ph.D. thesis. University of Wisconsin—Madison, 280 p.

Geochronology of the main deposits of the Penokean Orogeny of northeastern Wisconsin

Steven D.J. Baumann

Geochronology of the main deposits of the Penokean Orogeny of northeastern Wisconsin

Publication

G-112019-2A

Steven D.J. Baumann

First edition

November 2019

Midwest Institute of Geosciences and Engineering

10 Years (2009-2019)

Photo on the previous page was taken by Steven Baumann, taken on June 29, 2019 of "Twelve Foot Falls, Wisconsin".

Abstract

Most of the phaneritic plutons of the Penokean Orogeny, and related deposits, of the Wausau-Pembine Terrane have been either directly or relatively dated. We now have enough data to establish a pretty solid geochronology on these rocks.

The Wausau-Pembine is a an island arc accretion terrane that met up with the Superior Craton from about 1895-1822Ma (Schulz and Cannon, 2007). The initial lavas and phaneritic plutons of the Wausau-Pembine were emplaced in a short window of about 1880-1864Ma and are represented by the Quinnesec Formation and diorite as an island arc met the Superior Craton during subduction. The magmatic intrusions were emplaced relatively quickly in a period from about 1862-1835Ma and intruded the existing Quinnesec and Paleoproterozoic Chocolay Group. All magmatic activity appears to have happened within a narrow 45 million year window 1880-1835Ma.

Ages

Many of the phaneritic and aphanitic rocks have been dated. However, many are close together with relatively large margins of error. So where available, the law of cross-cutting relationships needs to be used to establish relative age. Combined with actual dates, and staying within margins of error, likely ages can be established. Below is a summary of the units and suspected ages based on dates and field relationships (see p. 3 for more information).

Unit	Age in Ma	Notes
Amberg Granite	1757	Post dates the Penokean Orogeny
Spikehorn and Bush Lake Granites	1835	Youngest known plutons and they're coeval
Athelstane Quartz Monzonite	1836/1839	This might actually be two plutons
Hoskin Lake Granite	1840 ?	2nd highest margin of error could be 1850Ma
Marinette Quartz Diorite	1859	The 1993 date has a small margin of error
Newingham Tonalite	1860	Emplacement age, not later altered age, highest margin of error
Dunbar Gneiss	1862	Emplacement age, not metamorphic age
Twelve Foot Falls Quartz diorite	1863 ?	Suspected emplacement age
Quinnesec Formation (undivided)	1864-1875	Includes meta-volcanics and meta-sediments
Quinnesec Diorite	1875-1880	Complex basal volcanic and plutonic series

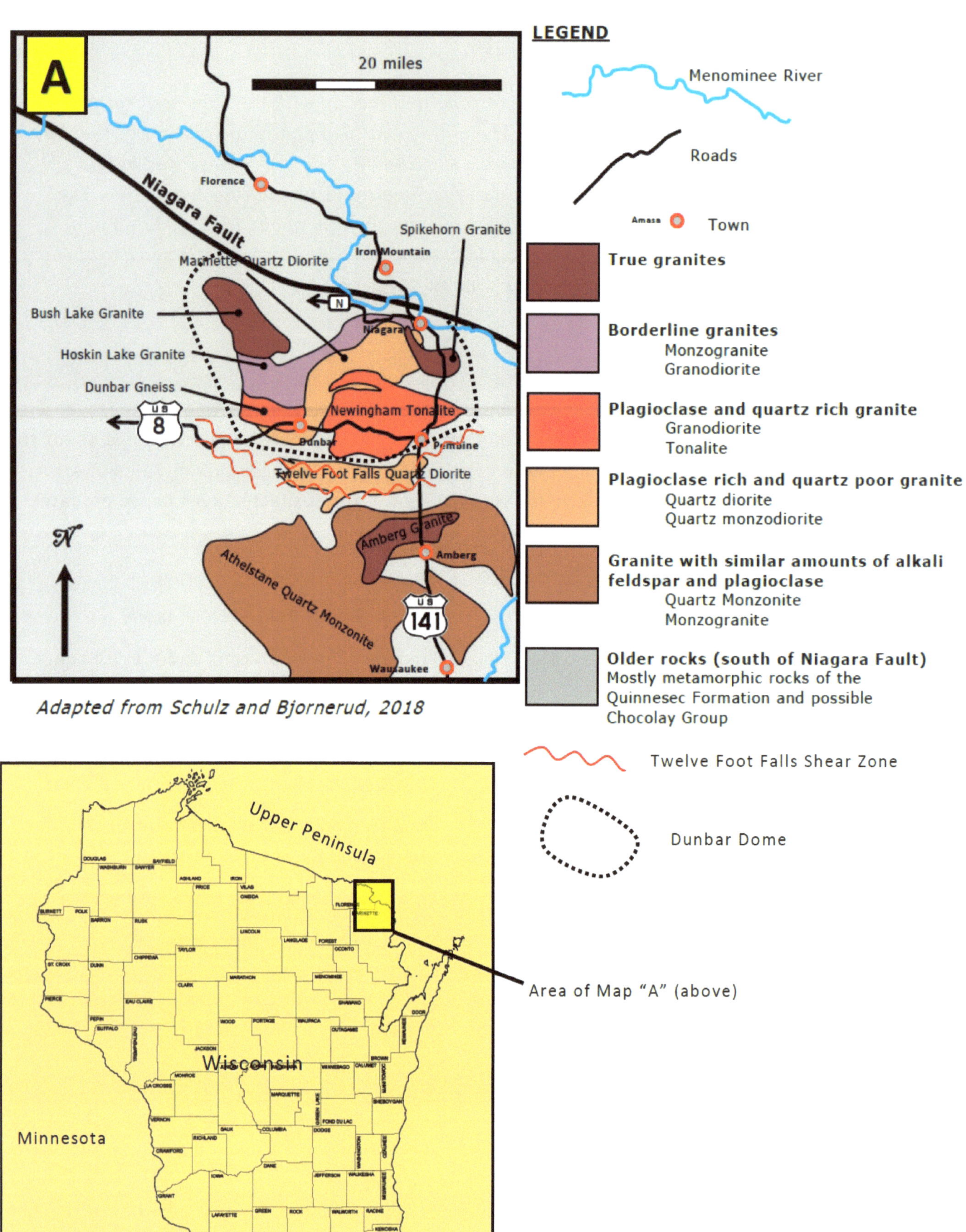

Adapted from Schulz and Bjornerud, 2018

Wausau-Pembine Terrane correlation chart

Age in Millions of Years Ago (Ma)	Geologic Symbol	Unit	(Dated age in Ma)	*Source for Age*	Notes
1100 1485	Xdd	Diabase dikes	(Possibly Keweenaw)		
1750 1760	Xag	Amberg Granite	(1759±19) (1754±11)	*Peterman et al., 1985* *Holm et al., 2005*	
1835	* Xsg	Spikehorn Creek Granite	(1836±6)	*Peterman et al., 1985*	Spikehorn Creek and Bush Lake are coeval
	* Xbg	Bush Lake Granite	(1835±5)	*Sims, 1990*	
	Xaq	Athelstane Quartz Monzonite	(1836±15)	*Sims, 1990*	
1840	* Xhg	Hoskin Lake Granite	(1861±32)	*Sims, 1990*	Hoskin Lake intrudes the Marinette
1855	* Xmd	Marinette Quartz Diorite	(1857+6) (1862±15)	*Sims and Schulz, 1993* *Sims, et al., 1992*	Marinette intrudes the Dunbar, Quinnesec, and Newingham
1860	* Xnt	Newingham Tonalite	(1861±40)	*Sims, 1990*	
1862	* Xdg	Dunbar Gneiss	(1862±4) (1862±5)	*Peterman et al., 1985* *Sims, 1990*	
1863 ?	Xtd	Twelve Foot Falls Quartz Diorite	(Not dated)		
1864	Xqu	Undivided Quinnesec Formation	(Not dated)		
1880	Xqd	Quinnesec Diorite	(1866±39)	*Peterman et al., 1985*	Age of the diorite was obtained from a rhyolite. Although it is called a dacite it is heterogeneous, consisting of dacite, rhyolite, and andesite.
2200 2450	Xcg	Chocolay Group			

Legend:

- Post Penokean magmatic activity
- Late Penokean magmatic activity
- Early Penokean magmatic activity
- Pre-Penokean rocks
- * Part of the Dunbar Dome

Geochronology of events during the Penokean Orogeny that formed the Wausau Pembine Terrane in northeastern Wisconsin

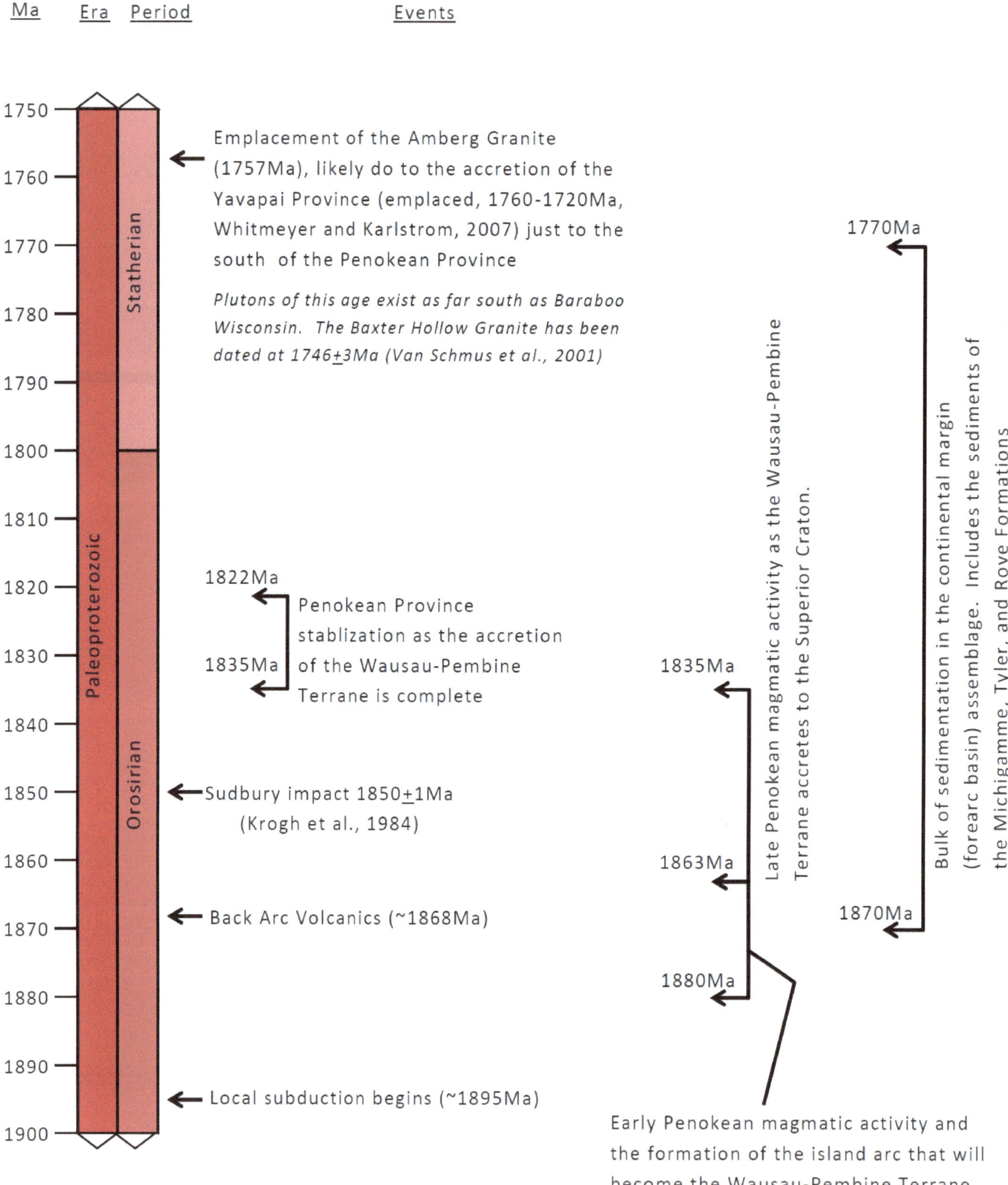

Small fault in the undivided Quinnesec Formation along US-141

GPS: 45.86620, -88.07803

Sample: 042119-1

U.S. penny = 0.75 inches (1.9cm) diameter

Spikehorn Granite along US-141

GPS: 45.72355, -87.94912

Sample: 042119-2 and 042119-4

U.S. penny = 0.75 inches (1.9cm) diameter

Amberg Granite along US-141

GPS: 45.48146, -87.98627

Sample: 060819-2

Engineering tape in inches and tenths of a foot.

1 inch = 2.54cm

1/10 of a foot = 3.05 cm

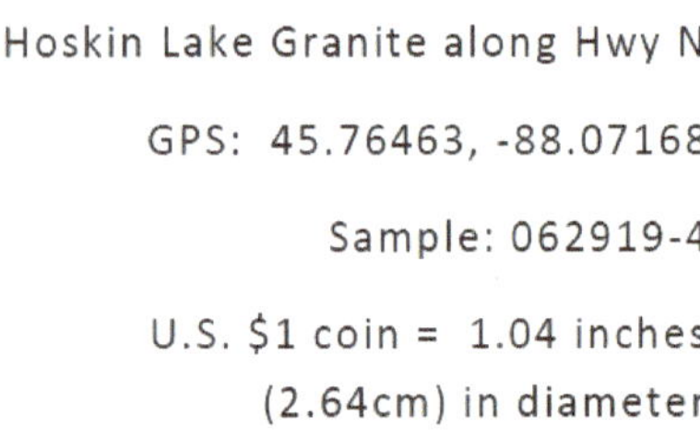

Hoskin Lake Granite along Hwy N

GPS: 45.76463, -88.07168

Sample: 062919-4

U.S. $1 coin = 1.04 inches
(2.64cm) in diameter

Megacryst facies of the Newingham Tonalite along US-8

GPS: 45.62268, -88.02359

Sample: 060819-2

Hammer is 11 inches (27.9cm) long

Amberg Granite along US-141

GPS: 45.46565, -87.98218

Sample: 071919-1

1/10 of a foot = 3.05 cm

Marienette Quartz Diorite off CC
Camp Road

GPS: 45.63889, -88.18990

Sample: 062919-7

U.S. $1 coin = 1.04 inches (2.64cm)
in diameter

Close-up of Athelstane Quartz
Monzonite (rapakivi type
texture) along US-141

GPS: 45.49552, -87.98427

Sample: 071919-2

1cm = 0.39 inches

Baumann, S.D.J., 2019. QAP plots of granitic rocks of northeastern Wisconsin. Midwest Institute of Geosciences and Engineering, publication number G 11 2019 1A

Holm, D.K., Van Schmus, W.R., MacNeill, L.C., Boerboom, T.J., Schweitzer, D., and Schneider, D., 2005. U-Pb zircon geochronology of the Paleoproterozoic plutons from the northern midcontinent, USA: Evidence for subduction flip and continued convergence after Geon 18 Penokean Orogenesis. Geological Society of America, Bulletin 2005, No. 117, Vol. 3-4, p. 259-279

Krog, T.E., Davis, D.W., and Corfu, F., 1984. Precise U-Pb zircon and baddeleyite ages for the Sudbury area. *in* Pry, E,G., et al., (editors). The geology and ore deposits of the Sudbury Structure. Ontario Geological Survey, Special Volume 1, p. 431-446

Peterman, Z.E., Sims, P.K., Zartman, and R.E., Schulz, K.J., 1985. Middle Proterozoic uplift events in the Dunbar Dome of northeastern Wisconsin USA.

Sims, P.K., 1990. Geologic map of Iron Mountain and Escanaba I°x2°quadrangles, northeastern Wisconsin and northwestern Michigan. U.S. Geological Survey Miscellaneous Investigations Series Map, I-2056

Sims, P.K., Schulz, K.J., and Peterman, Z.E., 1992. Geology and geochemistry of the Early Proterozoic rocks in the Dunbar area, northeastern Wisconsin. United Stated Geological Survey, Professional Paper 1517

Sims, P.K., and Schulz, K.J., 1993. Geologic map of Precambrian rocks of parts of Iron Mountain and Escanaba 30'x60' quadrangles, northeastern Wisconsin and adjacent Michigan. United States. Geological Survey Miscellaneous Investigations Series Map, I-2356

Schulz, K.J. and Cannon, W.F., 2007. The Penokean Orogeny in the Lake Superior Region. Precambrian Research, vol. 157, p. 4-25

Van Schmus, W. R.; MacNeill, L. C.; Holm, D. K.; and Boerboom, T. J. 2001. New U-Pb ages from Minnesota, Michigan, and Wisconsin: implications for Late Paleoproterozoic crustal stabilization. Institute on Lake Superior Geology Proceedings, 47th annual meeting (Madison, Wis.), 47, (abstract).

Whitmeyer, S.J. and Karlstrom, K.F., 2007. Tectonic model for the Proterozoic growth of North America. Geosphere, vol. 3, no. 4, p. 220-259. doi: 10.1130?GES00055.1

G-012020-1A(P): Chronostratigraphic nomenclature and geo correlation of the Precambrian and its map symbols

Printable pamphlet version

Compiled by: Steven D.J. Baumann (January 2020)

The far right column (map nomenclature), are the old symbols you would see on geologic maps that were set by the FGDC (see references). They don't exactly correspond to the Precambrian periods because they were meant to divide eons/eras into three divisions. There are many eons/eras with four periods/eras. As a result, they do not line up.

The Neoproterozoic Era's Cryogenian Period, was significantly reduced from the GSA 2012 (v4.0) timescale from a timespan of 215 million years; to the 2018 (v5.0) timescale with a span of only 85 million years (see references). This drastic modification is likely to change again.

This publication is my best attempt to reconcile the map nomenclatures with the timescale. It may not match up perfectly with some geologic maps (pre-2005).

Cover photo taken in August 2019 by Steven D.J. Baumann. Taken on Little Presque Isle, near Marquette, MI. It shows a columnar jointed Keeweenaw dike (Mesoproterozoic), cross cutting the local Archean gneiss.

REFERENCES:
Dalrymple, G.B., 1991: The age of the Earth, Stamford University Press, ISBN 10: 0804723311

Hofmann, H.J., 1990: Precambrian units and nomenclature-the geon concept, Geology (1990) 18(4):340-341

GSA Timescales, from various authors. The following versions were used: 1999, 2009, 2012 (v4.0), and 2018 (v5.0)

Unattributed authors, 2006, Federal Geographic Data Committee (FGDC): Digital cartographic standard for geologic map symbolization, FGDC Document number FGDC-STD-013-2006, prepared by the United States Geological Survey

G-012020-1A(P): Chronostratigraphic nomenclature and geo correlation of the Precambrian and its map symbols

Printable pamphlet version

Compiled by: Steven D.J. Baumann (January 2020)

Eon	Era	Period	Millions of years ago (Ma)	Age boundary pick (Ma)	Geon	Map Nomenclature

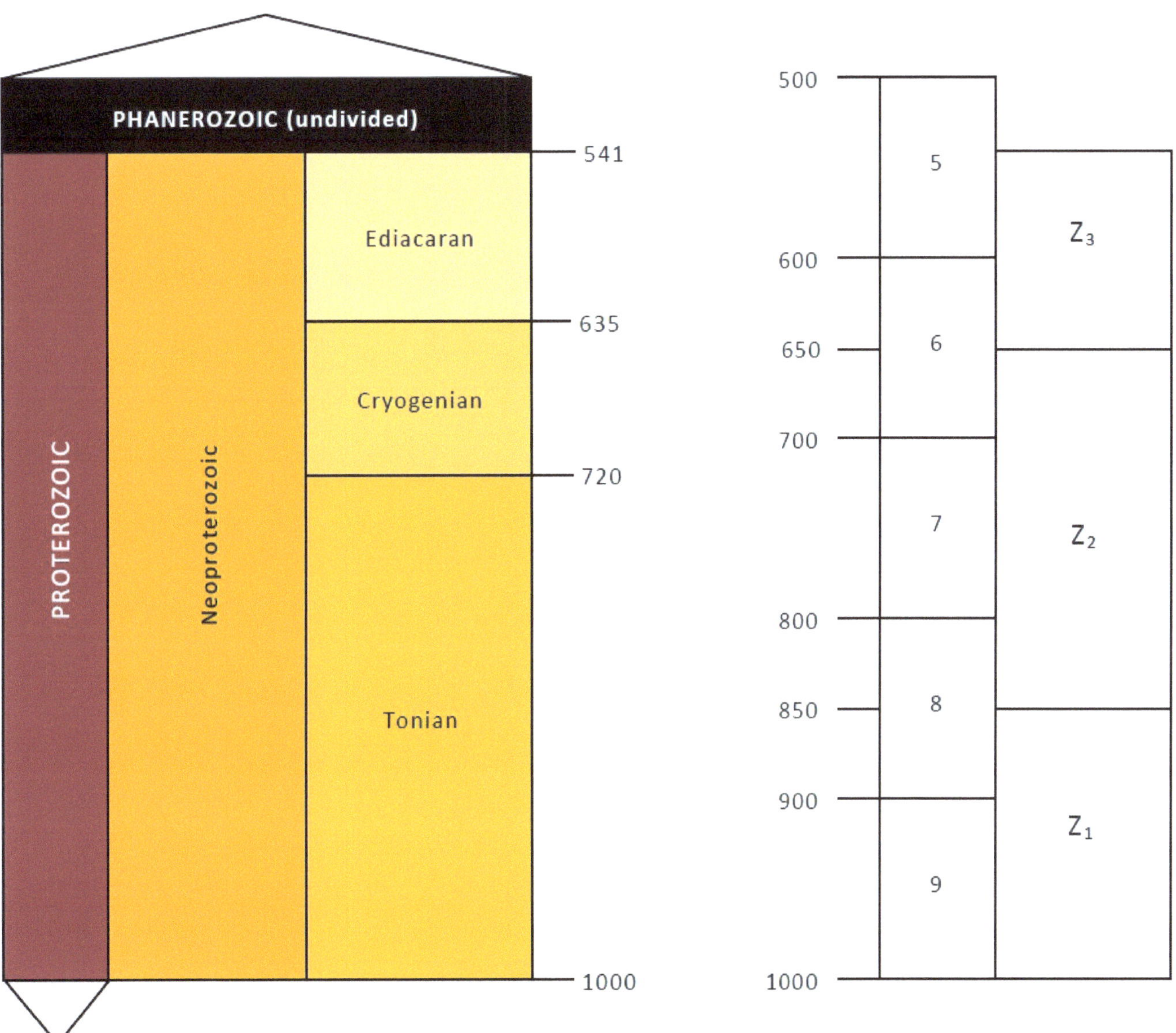

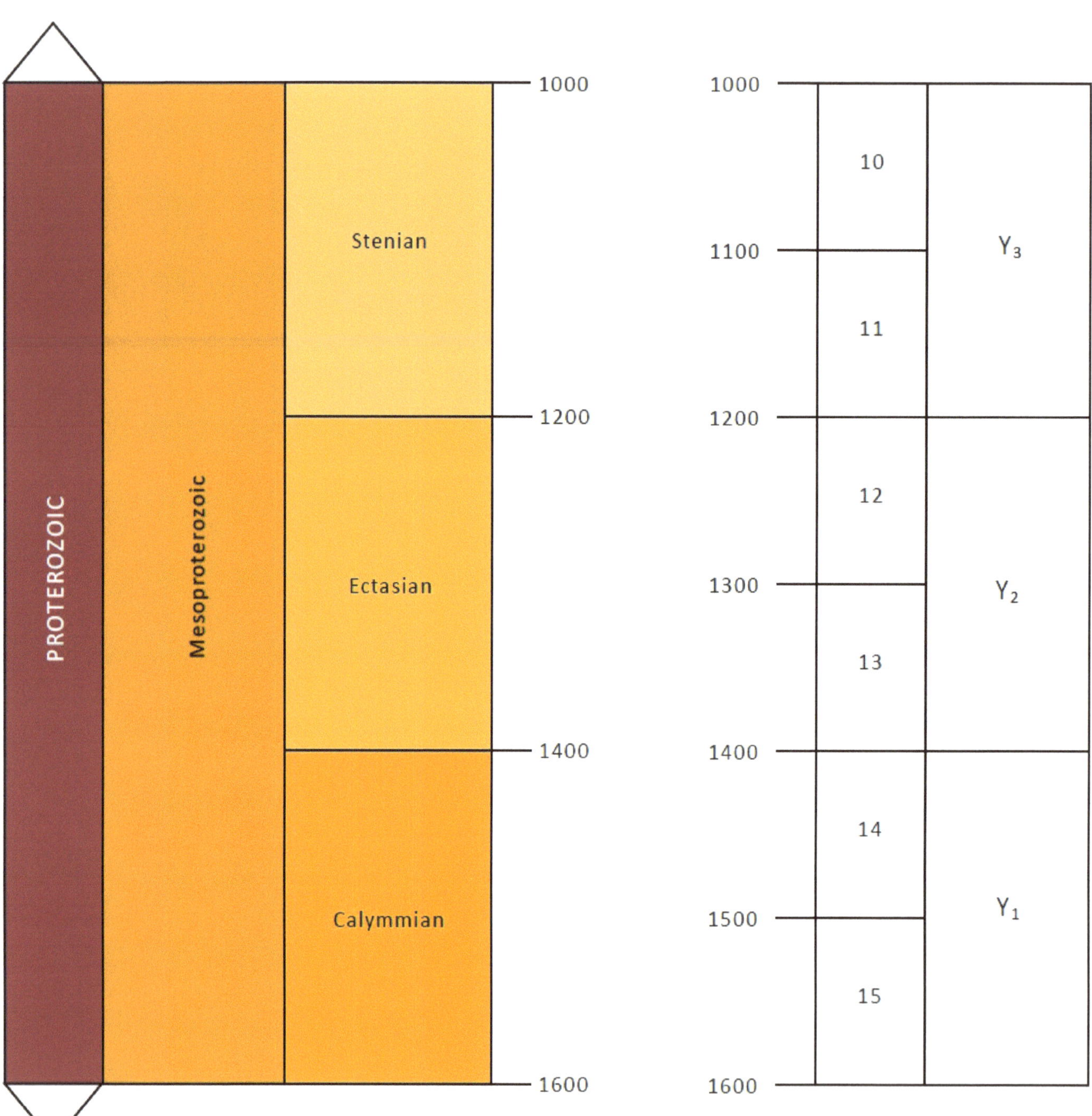

Eon
Era
Period
Millions of years ago (Ma)
Age boundary pick (Ma)
Geon
Map Nomenclature
PROTEROZOIC
Mesoproterozoic
Stenian
Ectasian
Calymmian
1000
1200
1400
1600
1000
1100
1200
1300
1400
1500
1600
10
11
12
13
14
15
Y3
Y2
Y1

Eon	Era	Period	Millions of years ago (Ma)	Age boundary pick (Ma)	Geon	Map Nomenclature
PROTEROZOIC	Paleoproterozoic	Statherian	1600	1600	16	
				1700	17	X_3
			1800	1800	18	
		Orosirian		1900	19	
				2000	20	X_2
			2050	2100	21	
		Rhyacian		2200	22	
			2300	2300	23	X_1
		Siderian		2400	24	
			2500	2500		

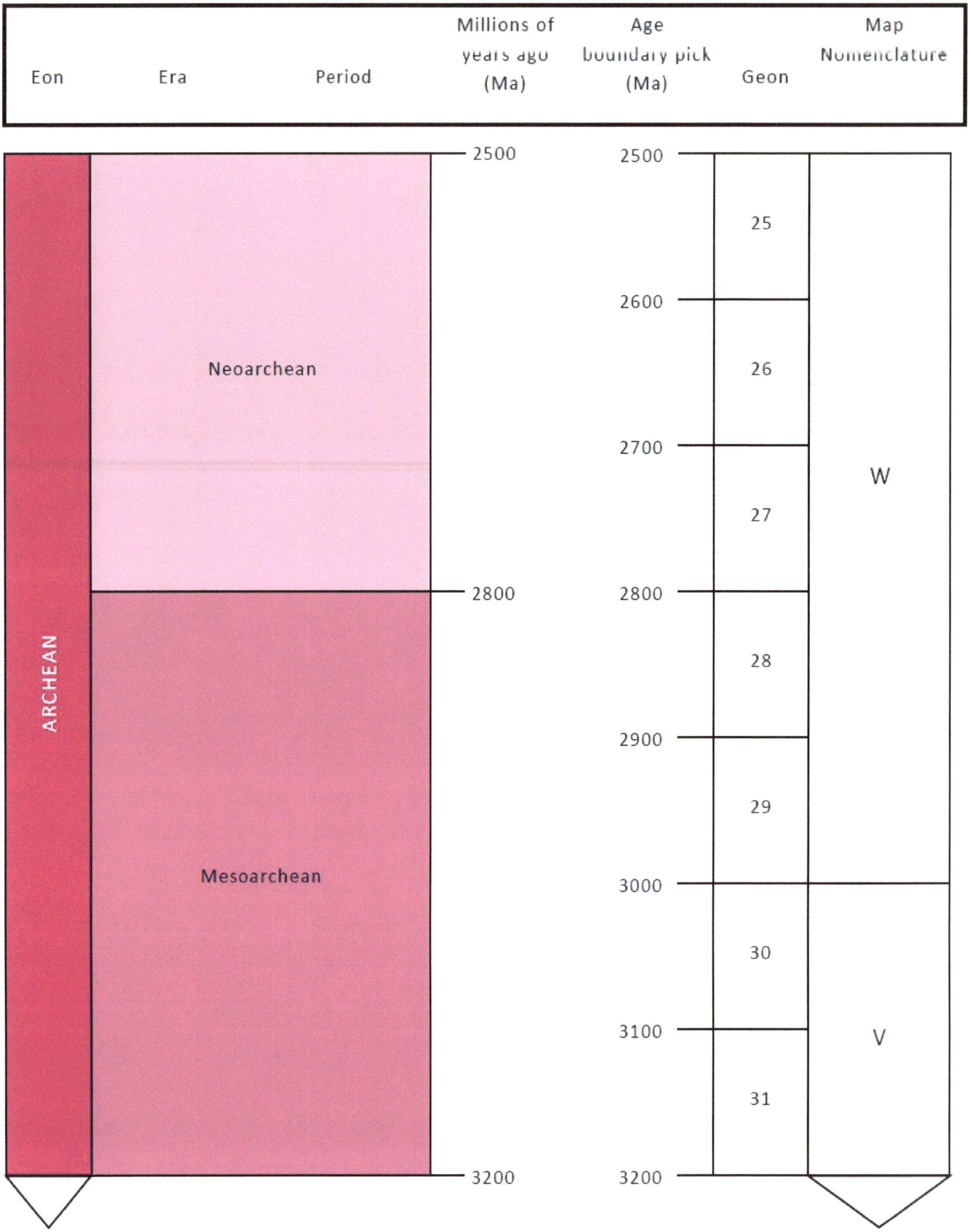

Eon
Era
Period
Millions of years ago (Ma)
Age boundary pick (Ma)
Geon
Map Nomenclature
ARCHEAN
Neoarchean
Mesoarchean
2500
2600
2700
2800
2900
3000
3100
3200
25
26
27
28
29
30
31
W
V

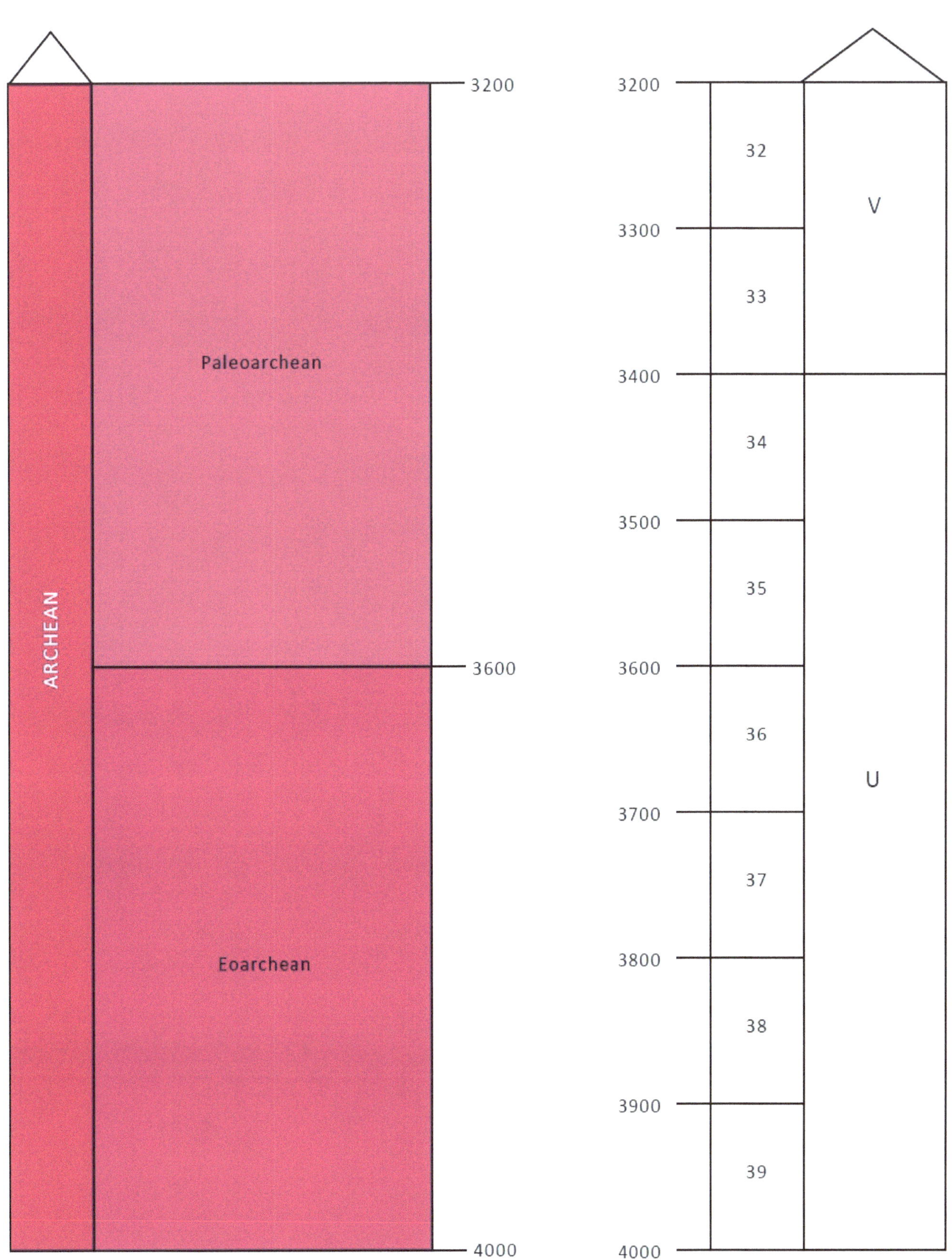

Eon
Era
Period
Millions of years ago (Ma)
Age boundary pick (Ma)
Geon
Map Nomenclature
ARCHEAN
Paleoarchean
Eoarchean
3200
3300
3400
3500
3600
3700
3800
3900
4000
3200
3300
3400
3500
3600
3700
3800
3900
4000
32
33
34
35
36
37
38
39
V
U

Eon	Era	Period	Millions of years ago (Ma)	Age boundary pick (Ma)	Geon	Map Nomenclature

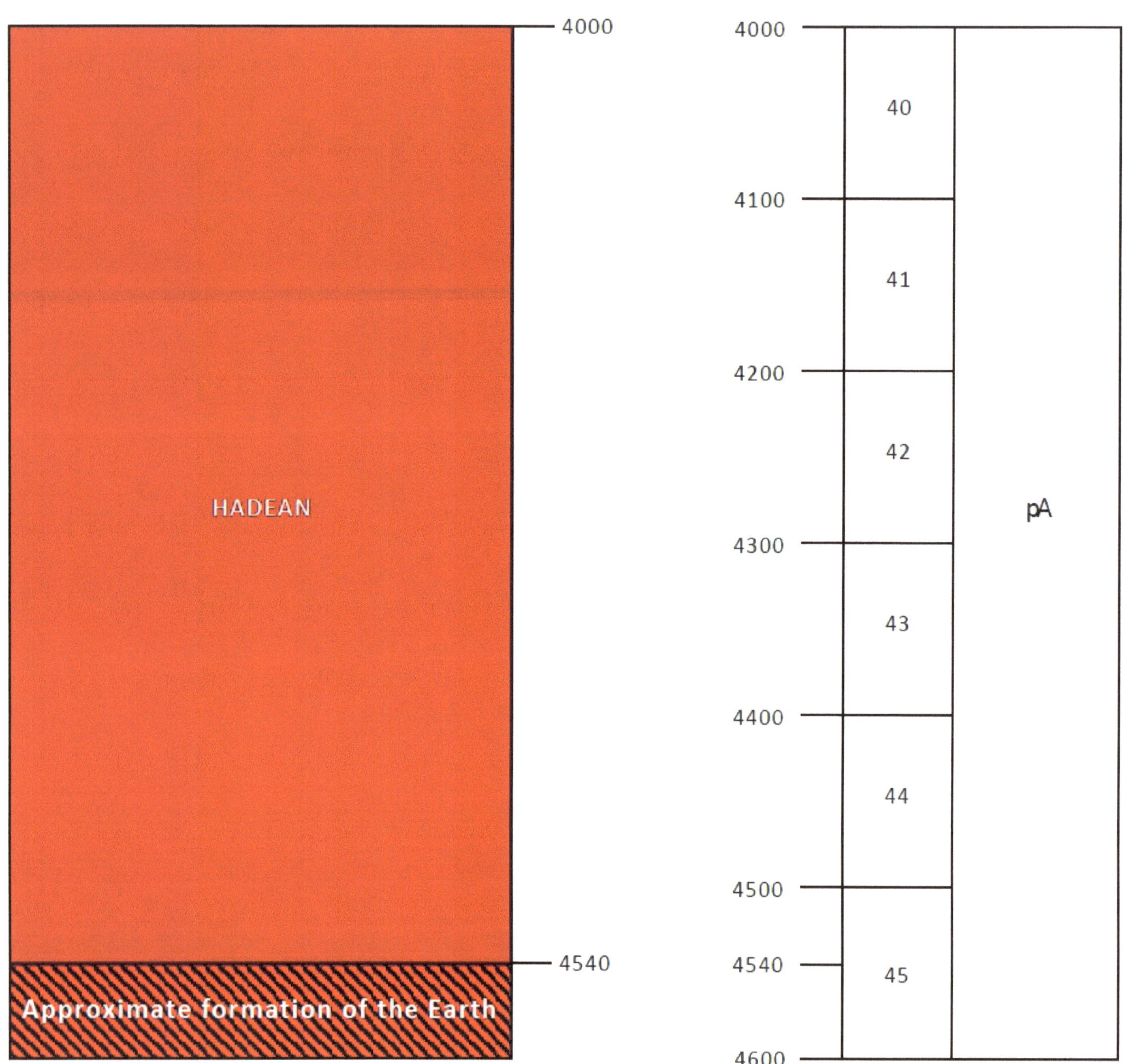

Possible quartzolite occurrence in the Michigamme Formation

Steven D.J. Baumann , Sarah M. Hall, Justin A. Meyers

Photo was taken on April 19, 2020by Steven D.J. Baumann at Calumet Waterworks Park up on the Keweenaw Peninsula looking northeast.

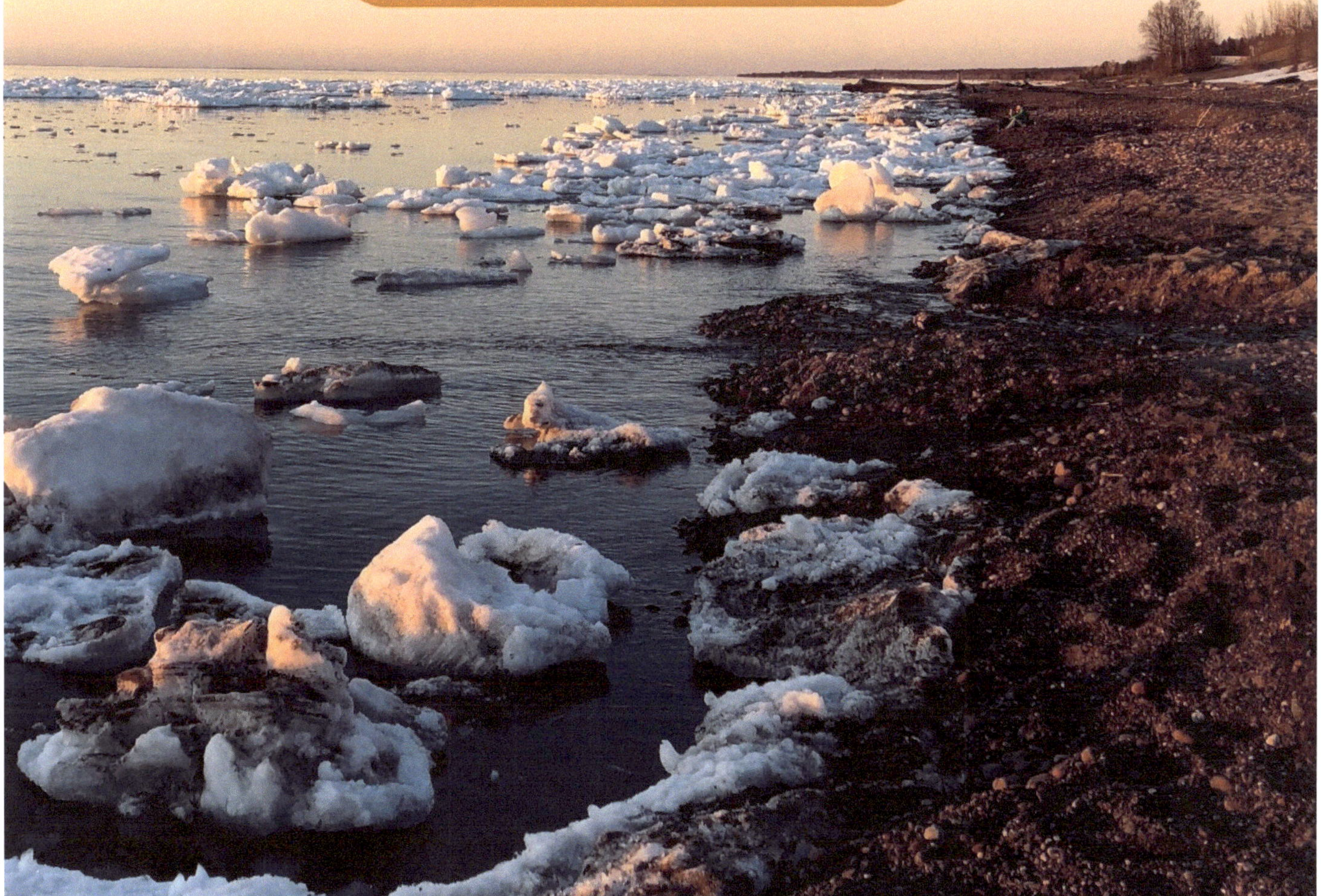

Possible quartzolite occurrence in the Michigamme Formation

Steven D.J. Baumann , Sarah M. Hall, Justin A. Meyers

There are several large quartz bodies throughout the Upper Peninsula of Michigan that cross cut the Michigamme Formation. It was assumed that these were just fracture filled veins or hydrothermal in origin (Seaman, 1947 and Waggoner, T., 2018). Some of the smaller ones definitely are, but not all. In the Negaunee quadrangle there are documented occurrences of a reddish brown granophyric quartz associated with diabase dikes (Puffett, 1974). Although this rock is closer to a quartz rich granitoid on a modern quartz-alkali feldspar-plagioclase-feldspathoid (QAPF) diagram (LeBas and Streckeisen, 1991).

There is a decent sized outcrop on the west side of US-141, about 1.35 miles (2.17km) northeast of Covington in Baraga County. The rock is composed of the greywacke and slates of the Michigamme Formation, referred to also as "host rock". Bedding varies from the base to the top of the outcrop but is about N60W52SW (middle) to N65W82SW (base). It is pretty typical Michigamme except for the large body of white rock near its center (figure 1 and 2). Six samples were taken for petrographic analysis, from the outcrop on December 16, 2018. Of these, three were taken from the white rock. Two of the three samples were quartz with weakly developed crystals. One sample (near the base of the white rock) was almost all muscovite. The muscovite makes up <5% of the volume of the white rock in outcrop and is confined to the bottom of the white rock. The rest of the white rock appears to be pure primary quartz.

According to Bowen's Reaction Series (Bowen, 1922), this is exactly what we would expect to see in a supersaturated magma body, not a hydrothermal vein. Muscovite would crystalize first. Due to the shape of the chamber, muscovite crystals would settle to the bottom first, leaving just quartz to crystalize. The muscovite is all at the bottom of the white rock as if it culminated out of magma, forming the quartzolite (figure 3). The quartz fills the rest of the body even into small offshoots which contain no muscovite. There is some altered green host rock disseminated throughout the bottom half, and into the small offshoots, of the white quartz body. The offshoots only contain quartz and incorporated host rock (figure 4). A hydrothermal deposit would not be so segregated. The white rock deposit has to be a quartzolite, cooled directly from a low temperature (perhaps <850°C), supersaturated magma body.

There are other large bodies of quartz that dwarf the one on US-141. Horse Race Rapids boasts a very large body of quartz at Quartz Mount, that may also be a quartzolite (Baumann, et al., 2016), but has not been studied in great detail.

The age of the quartzolite is not known but likely coincides with the Wolf River Event (~1450Ma) or the Midcontinent Rift (~1100Ma).

Figure 1: Outcrop, looking northwest, Steven D.J. Baumann for scale.

Figure 2: Approximate "to scale" diagram of the quartzolite outcrop

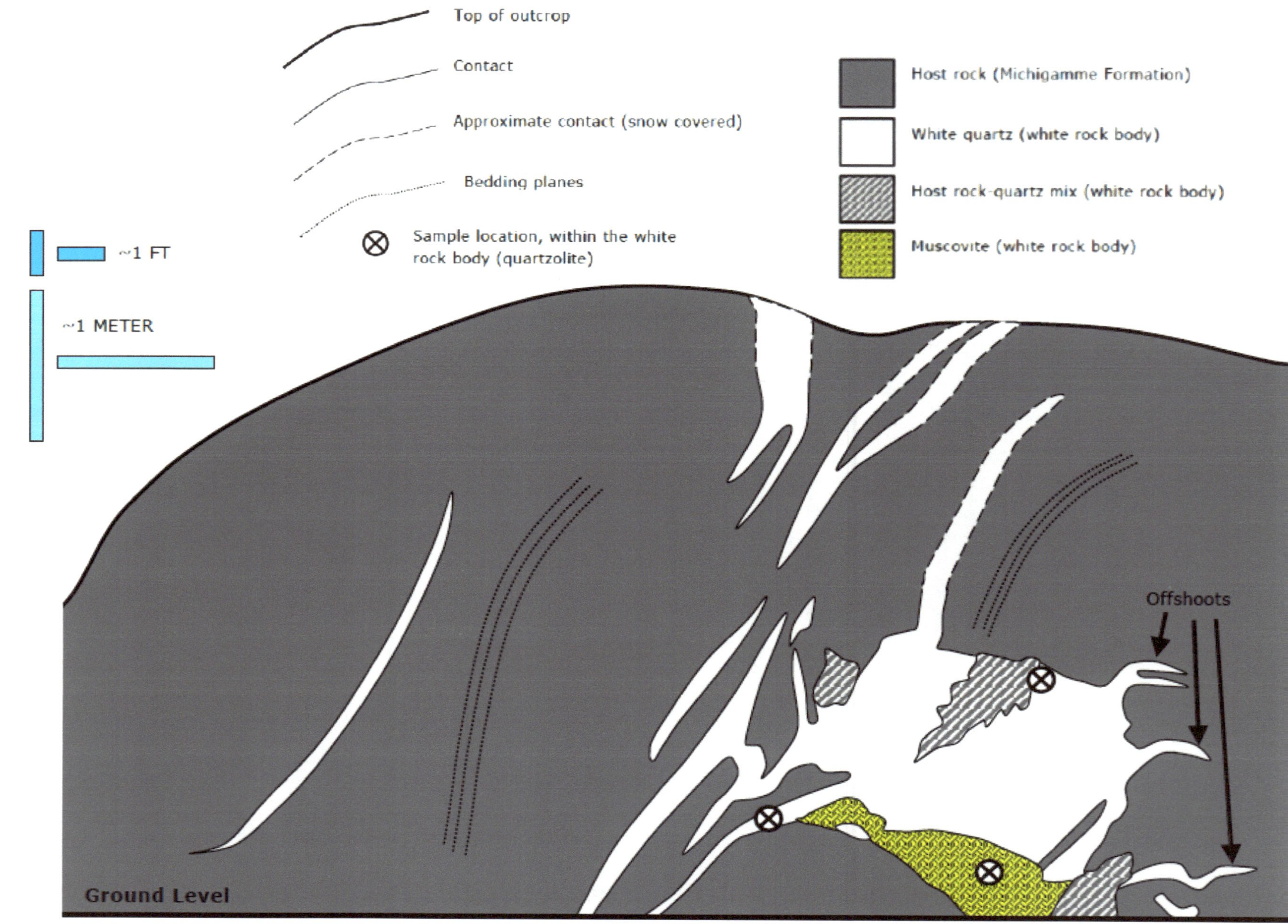

Figure 3: Generalized cross section of the formation of the quartzolite

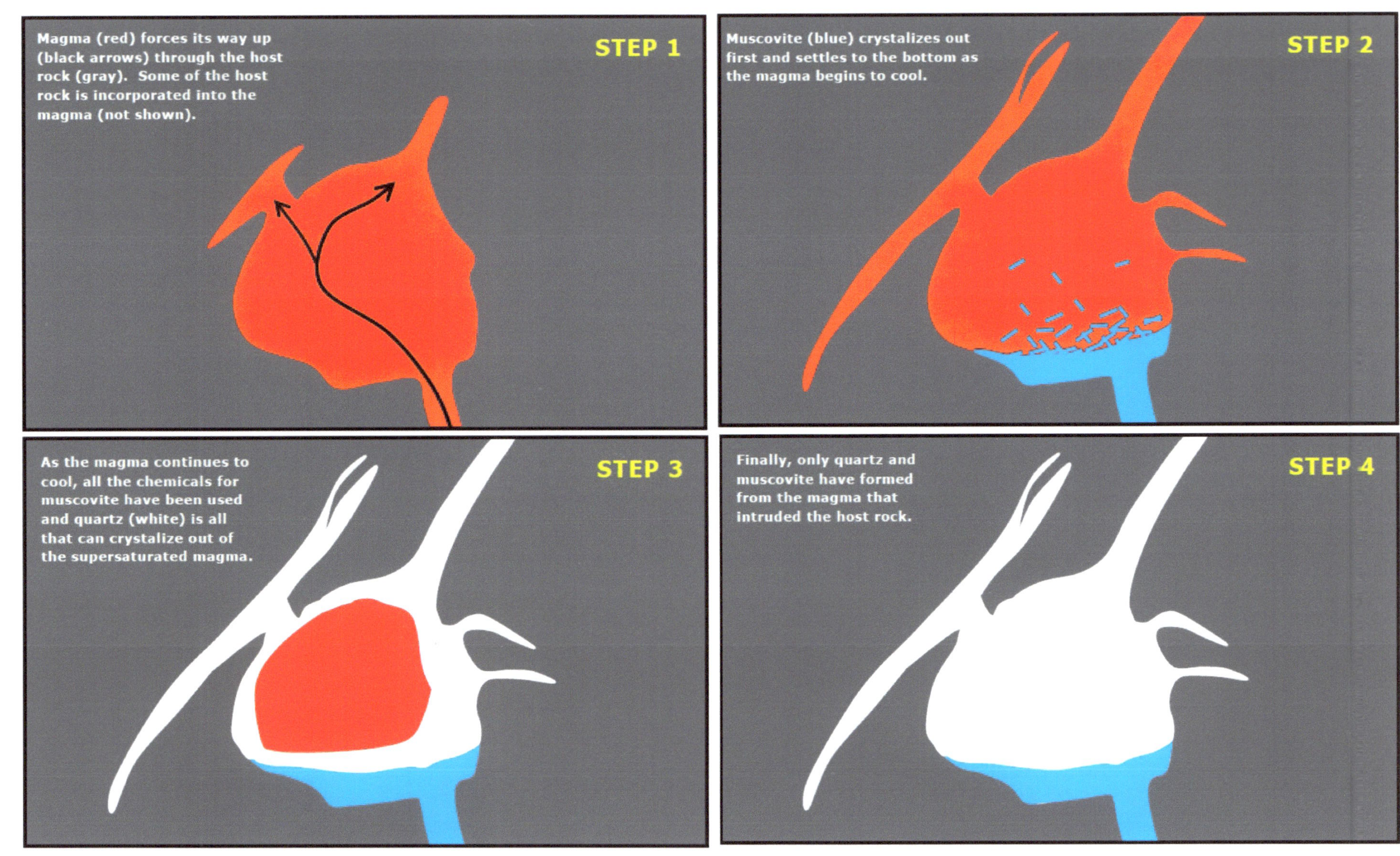

References

Baumann, S.D.J., Pryjda, M., and Johnson, D., 2016. Quartz Mount at Horse Race Rapids. Midwest Institute of Geosciences and Engineering, publication M-102016-2A

Bowen, N.L., 1922. The reaction principle in petrogenesis. Geology, v. XXX(3): 177-198

LeBas, M.J., Streckeisen, A.L., 1991. The IUGS systematics of igneous rocks. Geological Society, v. 148: 825-833

Puffett, W.P., 1974. Geology of the Negaunee Quadrangle, Marquette County, Michigan. United States Geological Survey, Professional paper 788, 53 p.

Seaman, W.A., 1947. Geology of the Lake Michigamme area, Baraga and Marquette Counties. State of Michigan, Department of Conservation, Geological Survey Division, Open File Report 68, 38 p.

Waggoner, T. and Cannon, W.F. (compiler), 2018. Proceedings volume 64, Part 2: Field trip guidebooks. 64th annual meeting, Institute on Lake Superior Geology, Iron Mountain, Michigan, May 15-18, 2018, 50-51

Metamorphic Index Minerals, Facies, and Nodes

 Compiled by: Steven D.J. Baumann

Metamorphic Index Minerals, Facies, and Nodes

64

Publication Number: G-052020-1A

Compiled by: Steven D.J. Baumann

Midwest Institute of Geosciences and Engineering

First edition: May 2020

Photo on the previous page was taken at an outcrop in the Grenville Province in Ontario, just east of Sudbury on the north side of Trans-Canada 17.

Photo taken by: Steven D.J. Baumann, on November 24, 2018.

Abstract

This short publication is a compilation of may petrographic sources on metamorphic rocks (see references). It is meant to be used as a reference in basic petrographic analysis, coupled with the field relationships, of metamorphic rocks with a mudstone (or argillaceous rock) as their protolith. This publication is not a teaching guide!

PLATE 1:

Plate 1 is a comprehensive guide that shows different rock types and index minerals that form as a sedimentary rock becomes increasingly metamorphosed. I have included 9 common index minerals as well as several minerals that are not index minerals. Some minerals like quartz, some feldspars, calcite, and dolomite are stable from the onset of metamorphism to melting, so they are not metamorphic index minerals. As a mudstone, or other dominantly pelitic rock, becomes metamorphosed, certain minerals will form within it as increased heat and pressure changes the chemical arrangement and minerology of the rock (also see Plate 2) and new minerals will form from the clays as metamorphism increases.

PLATE 2:

Plate 2, gives basic data on the 9 index minerals in Plate 1, as well as 3 additional common minerals that are not included on Plate 1. There are also additional notes for each mineral.

PLATE 3:

Plate 3 is in two segments. Both contain the same background template of the metamorphic facies. It was split in two to avoid it being overcrowded. The top diagram has all the metamorphic facies named on it. Note that lithification/diagenesis is not a continuous yellow. That's because it is not a metamorphic facies. The top chart also contains andalusite, kyanite, and sillimanite index areas. These three index minerals are included because they are trimorphic with each other. Trimorphism means they all have the same chemical formula but crystallize in 3 distinct forms. None of the other index minerals do that.

The bottom chart traces the center lines of the track that a rock will follow under certain tectonic regimes. The plate gradient also includes the main types of rock as a mudstone progresses to a gneiss during metamorphism. This happens on the other tracks as well (contact, arc, and subduction gradients), they just aren't labeled as there is some variability and overlap.

PLATE 4:

Plate 4 deals with the concept of metamorphic nodes on a regional map scale. Nodes are the areas of metamorphism delineated by index mineral zones over a large area. Things like plutons may be within them, but they are not part of the node. The concept of nodes was developed mainly by James (1954-1955) while he was doing work in the Upper Peninsula of Michigan (see Plate 5). As rocks become metamorphosed it is rarely from a single point with one index mineral surrounding that point. The centers of nodes are often stretched out with different index minerals extending from the highest area of metamorphism to the least. This is because regional metamorphism (like during an orogeny) is extremely common.

PLATE 5:

Plate 5 shows the concept of metamorphic nodes in real space. It's an adaptation of the work by James in 1954-1955. Most of the metamorphism in the area took place during the Penokean orogeny (~1850Ma). Later the volcanics and clastics of the Midcontinent Rift (not part of any node) would cut through the area (~1100Ma). Even later the Paleozoic seas would transgress the area (~500Ma), which also are not part of any node. The surficial Quaternary unconsolidated cover does not appear on Plate 5.

Metamorphic Grades and Index Minerals for Metamorphic Rocks with a Sedimentary Protolith

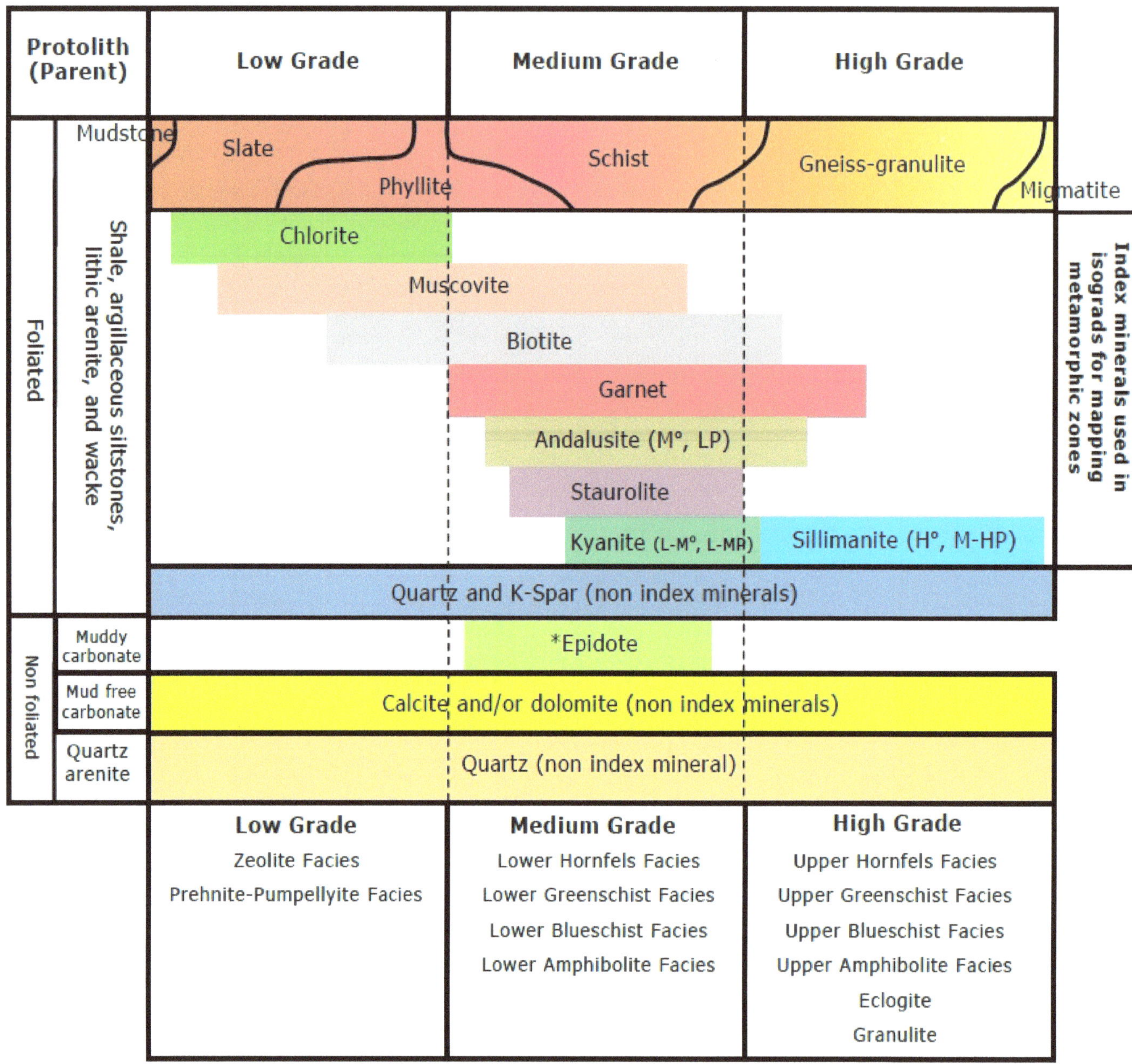

Some charts do not have a Prehnite-Pumpellyite Facies. They will instead be generally divided into a very low greenschist and low blueschist facies. Often referred to in older publications as just lower greenschist or lower blueschist facies.

The index minerals are generally only present in rocks whose origin, is at least in part, clayey. Not all index minerals will be present through all metamorphic isograds. Rock chemistry determines which index minerals will be formed.

(M°, LP) = Medium temperature, low pressure

(L-M°, L-MP) = Low to medium temperature, low to medium pressure

(H°, M-HP) = High temperature, medium to high pressure

* Epidote is also an index mineral in schists and occupies the same space as it does in muddy carbonates.

PLATE 1

Common Metamorphic Minerals that form in Argillaceous to Clay Protolith Rocks

Mineral	Chemical Formula & *Crystal System*	Notes
Andalusite	Al_2SiO_5 *Orthorhombic*	A neosilicate that is trimorphic with kyanite and sillimanite.
Biotite	$K(Mg,Fe)_3AlSi_3O_{10}(F,OH)_2$ *Monoclinic*	A dark colored mica. Usually black but it can be a dark golden brown or dark gray.
Chlorite	$(Mg,Fe)_3(Si,Al)_4O_{10}(OH)_2 \cdot$ $(Mg,Fe)_3(OH)_6$ *Monoclinic w/ triclinic polymorphs*	Chlorite is a group of 11 individual phyllosilicate minerals that cannot readily be individually identified in metamorphic rocks without detailed chemical analysis .
Epidote	$Ca_2Al_2(Fe^{3+};Al)(SiO_4)(Si_2O_7)O(OH)$ *Triclinic*	Epidote is a group of sorosilicates as well as an individual mineral. The information for the individual mineral is given herein. It forms exclusively as a secondary mineral mostly in hydrothermal, metamorphosed muddy carbonate, and sometimes in schists.
Garnet (Group)	$(Mg,Ca,Mn,Fe)_3(Al,Fe,Cr)_2Si_3O_{12}$ *Isometric and rarely tetragonal*	A group of 6 common neosilicate minerals with an additional 21 rarer minerals. Although nearly all garnets are isometric 2 very rare ones majorite and henritermierite are tetragonal. Garnets, unlike many metamorphic index minerals, will often form very distinct crystals in metamorphic rocks and will be translucent pink to opaque deep red in color.
Glaucophane	$Na_2(Mg_3Al_2)Si_8O_{22}(OH)_2$ *Monoclinic*	An inosilicate that along with lawsonite (not included herein) is common in the metamorphic blueschist facies and is usually a distinct bluish gray color.
Kyanite	Al_2SiO_5 *Triclinic*	A neosilicate that is trimorphic with andalusite and sillimanite.
Muscovite	$KAl_2(AlSi_3O_{10})(F,OH)_2$ *Monoclinic*	A light colored mica that is often called white mica. Although it can be other light colors.
Serpentine (Subgroup)	$(Mg,Fe)_3Si_2O_5(OH)_4$ *(variable)*	A subgroup of minerals usually associated with retrograde metamorphism.
Staurolite	$(Fe^{2+},Mg)Al_9(Si,Al)_4O_{20}(O,OH)_4$ *Monoclinic*	A neosilicate that is usually reddish brown and crystals will often form cross shapes, due to twinning at 60° and 90°.
Sillimanite	Al_2SiO_5 *Orthorhombic*	A neosilicate that is trimorphic with andalusite and kyanite.
Talc	$Mg_3Si_4O_{10}(OH)_2$ *Triclinic/Monoclinic*	A magnesium silicate that results from the metamorphism of magnesian minerals such (amphibole, olivine, pyroxene, serpentine) in the presence of carbon dioxide and water. It is the defining mineral of 1 on Mohs hardness scale and rarely forms crystals.

PLATE 2

Metamorphic Facies Diagrams

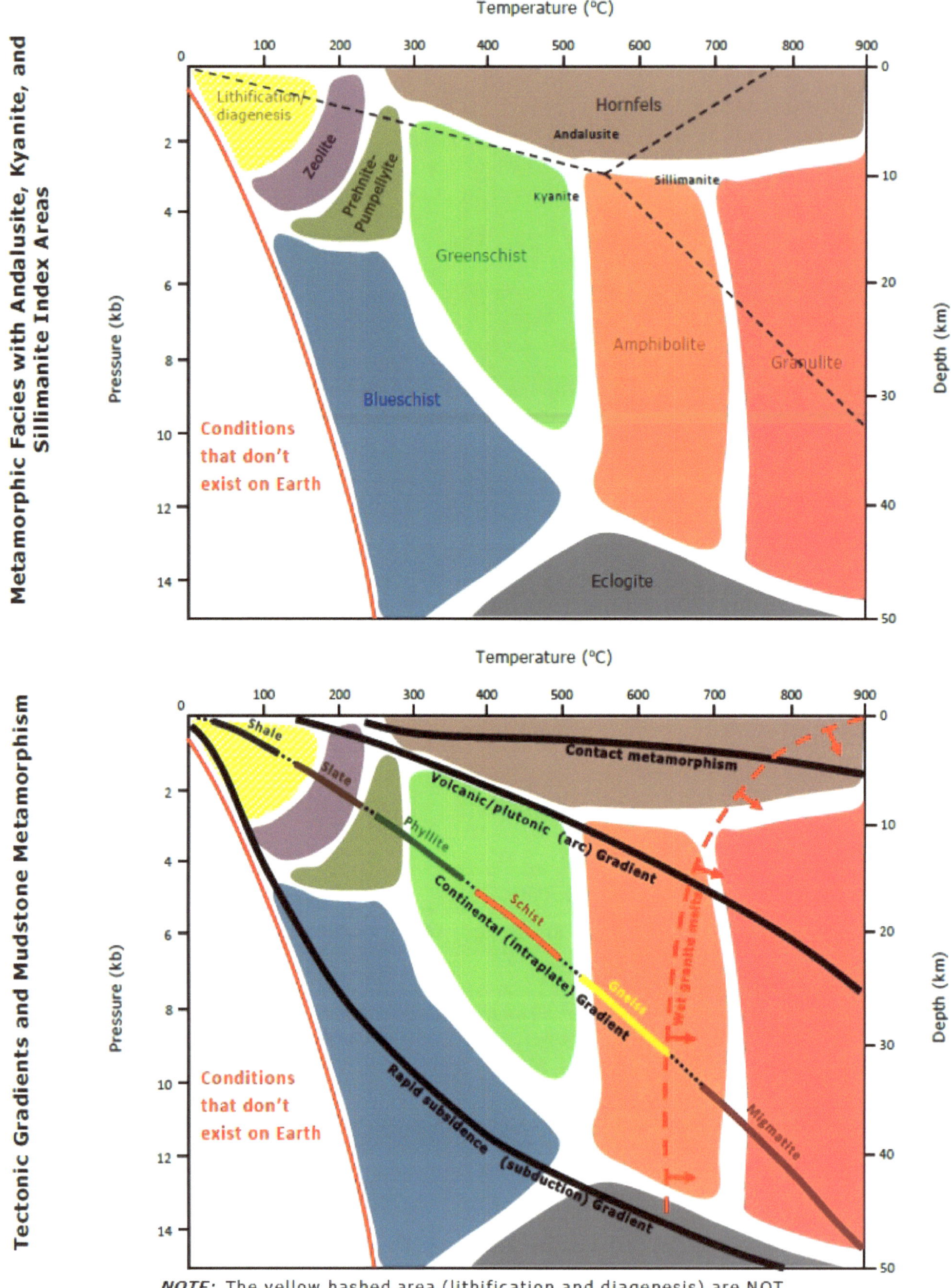

Metamorphic Facies with Andalusite, Kyanite, and Sillimanite Index Areas

Tectonic Gradients and Mudstone Metamorphism

NOTE: The yellow hashed area (lithification and diagenesis) are NOT metamorphic facies. Rocks in that area would be considered sedimentary.

PLATE 3

Hypothetical Illustration of Metamorphic Isograds and Nodes in Pelitic (clay rich) Rocks on a Regional Scale

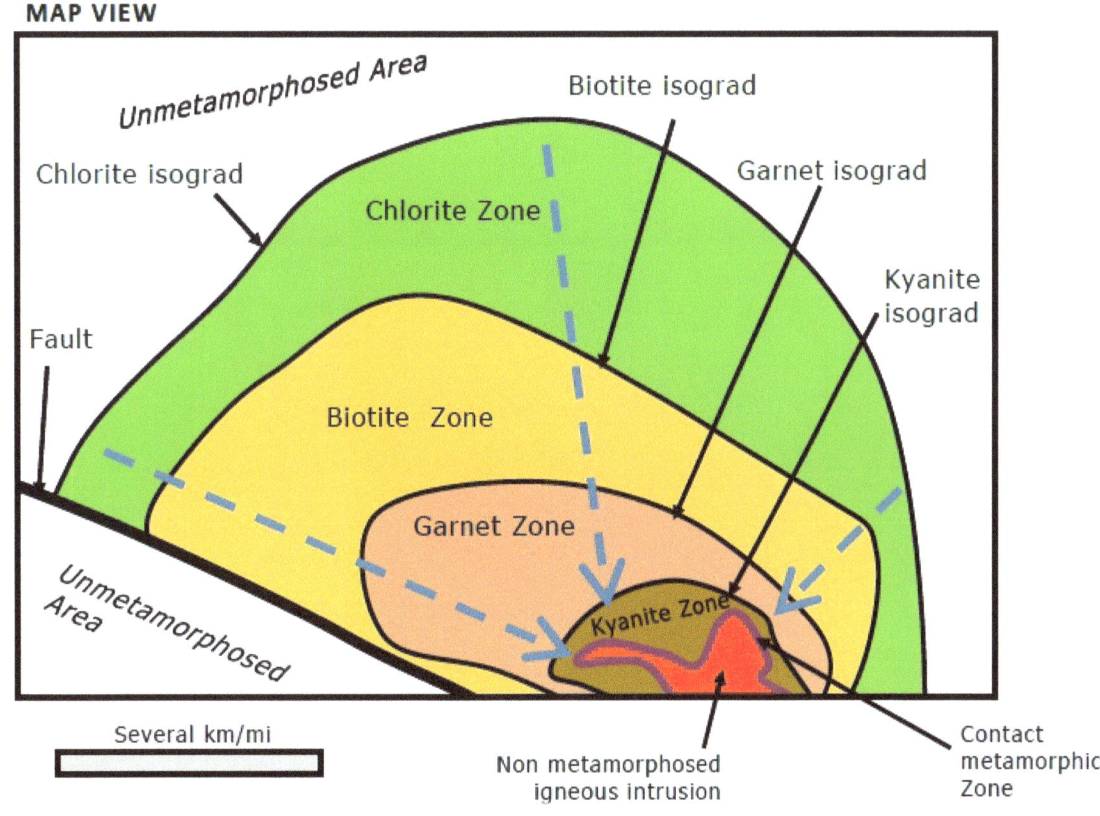

This is a hypothetical diagram of what a metamorphic node looks like. The "node" includes all of the rocks divided into "zones" and includes all the metamorphosed rocks within the outer isograd. The white area is not metamorphosed and is outside of the node. A zone is an area where a certain index mineral dominates. The zones are separated by isograds. Isograds are drawn at the first outer occurrence of an index mineral.

In the example above; one isograd is thicker than the rest (in this case, the purple line) and is the area immediately surrounding the red intrusion. This purple isograd is the zone where contact metamorphism occurred. It is the area of where we would find hornfels.

The dashed blue arrows just point to the direction of greater metamorphism. The blue arrows would not appear on a map, unless the scale of the map won't allow for the depiction of other zones.

Nodes can (and almost always will) cross lithological units.

In the above case the node contains no muscovite, andalusite, epidote, or staurolite zone because the rock chemistry favored the formation of chlorite, biotite, garnet, and kyanite. In the above example the process of metamorphism stopped in the medium grade, upper greenschist facies. So we wouldn't see any high grade index minerals like sillimanite.

The above node is bounded by the chlorite isograd and the fault. The igneous intrusion is not part of the node. The main metamorphism depicted could have been formed by the fault, the intrusion, or both.

Nodes vary in composition, shape, size, and extent. Although most are at least 260km^2 (or about 100mi^2) or larger in area.

In the above example, assuming all the metamorphic rocks have a mudstone protolith and follow a normal intraplate gradient; you would likely find the chlorite zone is made of slate and is of low grade (zeolite to prehnite-pumpellyite facies) metamorphism. The biotite zone is low to medium grade (prehnite-pumpellyite facies) and would likely be phyllite. The garnet zone would be schist and would represent medium grade (greenschist) facies. The inner kyanite zone is also medium grade (greenschist to amphibolite facies) but it forms after the garnet facies and would likely be a combination of schist and gneiss.

PLATE 4

Metamorphic Nodes (Zones) in the Upper Peninsula of Michigan

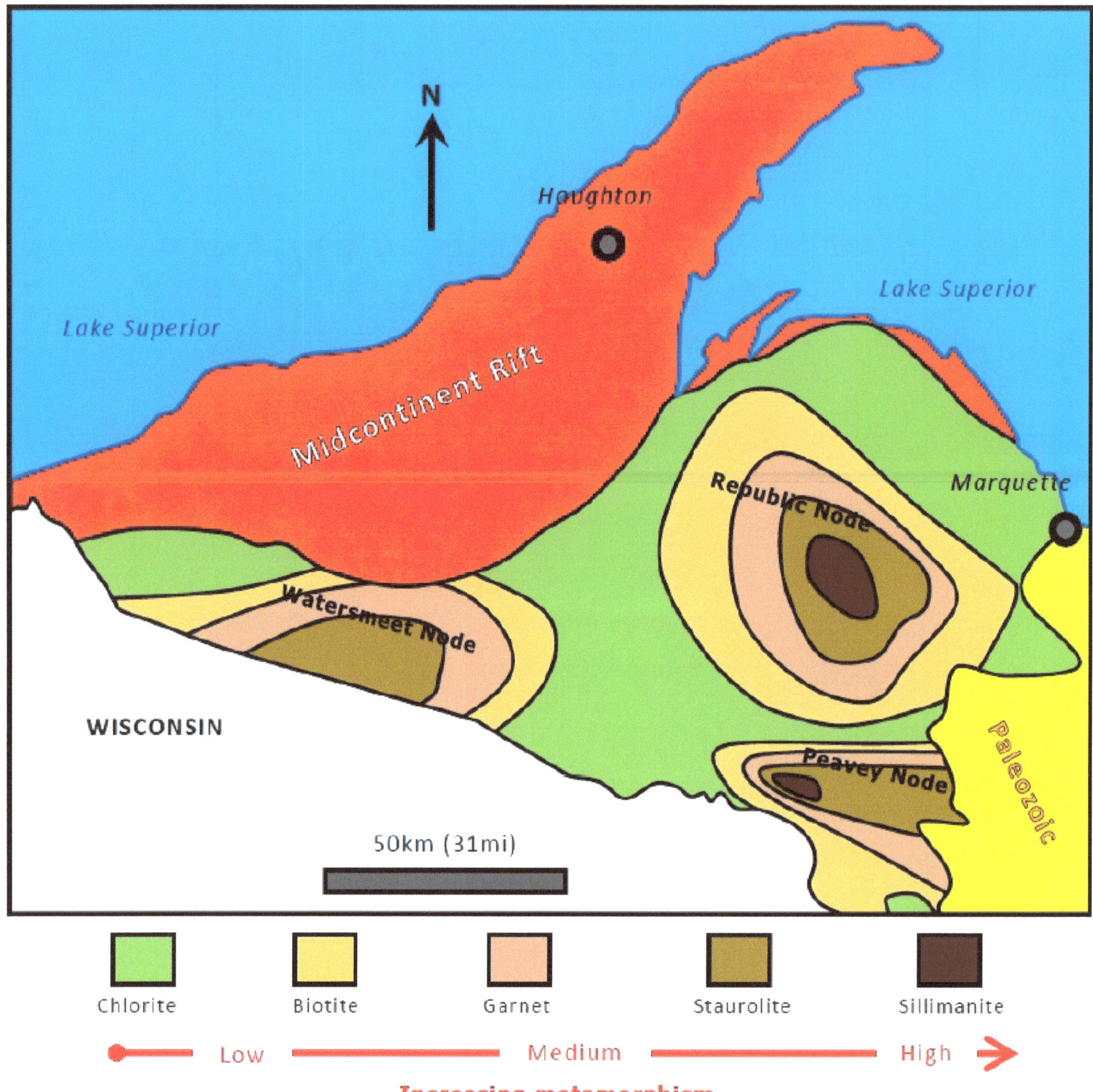

The above map shows the named metamorphic nodes that occur in the Upper Peninsula of Michigan. The Midcontinent Rift (MCR) and Paleozoic are outside of the 3 metamorphic nodes. The 3 nodes are the Peavey Node, Republic Node, and Watersmeet Node. The MCR and Paleozoic deposits are not part of any metamorphic nodes. They were not formed by orogenic events.

PLATE 5

References

Anthony, J.W., Bideaux, R.A., Bladh, K.W., and Nichols, M.C. (editors), 2004-2020. Handbook of Mineralogy. Mineralogical Society of America, handbookmineralogy.org

Bucher, K. and Grapes, R., 2011. Petrogenesis of Metamorphic Rocks. ISBN: 9783540741688

Bushmin, S.A., and Glebovitsky, V.A., 2007. Scheme of mineral facies of metamorphic rocks. Institute of Precambrian Geology, Russian Academy of Sciences, DOI: 10.1134/S1075701508080011

Coombs, D.S., 1989. Prehnite-pumpellyite facies. Encyclopedia of Earth science. Springer, Boston, MA, DOI: 10.1007/0-387-30845-8_200

Faust, G.T. and Fahey, J.J., 1962. The serpentine-group minerals: studies of the natural phases in the system MgO-SiO_2-H_2O and the systems containing the congeners of magnesium. United Stated Geological Survey, Professional Paper 384-A,

Garlick, G.., 1965. Oxygen isotope ratios in coexisting minerals of regionally metamorphosed rocks. PhD thesis, California institute of technology

Grotzinger, J. and Jordan, T.H., 2014. Understanding Earth. 7th edition, published by: W.H. Freeman and Company, ISBN: 9781464138744

James, H.L., 1954-1955. Zones of regional metamorphism in the Precambrian of Northern Michigan. Geological Society of America Bulletin, v. 66, p.1455-1488, plate 1

Marshak, S., 2015. Earth (Portrait of a Planet). 5th edition, Chapter 8, ISBN: 9780393937503

Nelson, S.A., 2011. Metamorphic reactions, isograds, and reaction mechanisms. EENS 212 lecture notes, Tulane University

Robertson, S., 1999. BGS rock classification scheme, classification of metamorphic rocks. British Geological Survey Report, RR 99-02, vol. 2

Chronology and redefinition of the Mesoproterozoic, Paleoproterozoic, Archean, and Hadean

Chronology of the Mesoproterozoic, Paleoproterozoic, Archean, and Hadean

This photo log corresponds to the photos on the previous page.

Glacier National Park
Grinnell Trail
Grinnell Formation
Looking south

Taken on: September 7, 2019
Taken by: Steven D.J. Baumann

Glacier National Park
Grinnell Trail
Appekunny Formation
Looking northwest

Taken on: September 7, 2019
Taken by: Steven D.J. Baumann

Glacier National Park
Grinnell Trail
Grinnell Formation
Looking east

Taken on: September 7, 2019
Taken by: Steven D.J. Baumann

Glacier National Park
Grinnell Trail
Appekunny Formation
Looking west

Taken on: September 7, 2019
Taken by: Steven D.J. Baumann

Glacier National Park
Iceberg Trail
Grinnell Formation
Looking Southwest

Taken on: September 6, 2019
Taken by: Steven D.J. Baumann
Background: Sarah M. Hall

Chronology of the Mesoproterozoic, Paleoproterozoic, Archean, and Hadean

ABSTRACT

The Precambrian geologic time scale has traditionally been largely very loosely based off of a handful of absolute dates. It is largely arbitrary. Herein, we continue the tradition of absolute dating as the borders of redefined and/or new epochs, periods, and eras in the Precambrian. Absolute dating of selected igneous events or major cratonic events or impacts, have to be used instead of fossils; which are commonly used in the Phanerozoic. The reason being, is that in the field and even in some lab settings (other than stromatolites) fossils aren't readily identifiable in the Precambrian, if they exist at all. We redefine the Hadean (and formalize it), Archean, Paleoproterozoic, and Mesoproterozoic. The Neoproterozoic is excluded due to its glacial complexity and the emergence of complex life. We now have really good Precambrian dates that we didn't even have in the early 2000's. Now, a detailed division of the Precambrian is possible and arbitrary boundaries are no longer needed. The Hadean is divided into three eras and several periods. The Archean is redefined and also divided into redefined eras and periods. The Hadean and Archean are not further divided into epochs. For the first time an attempt is made to divide the Paleoproterozoic and Mesoproterozoic into epochs, with a couple of exceptions. Some of the period names that appear in the "Geological Society of America's" time scale have been redefined or abandoned altogether.

INTRODUCTION

The term "Precambrian" is not a formal division of geologic time. It's a generic term used for all of Earth's history before the Phanerozoic. Which is presently set at 541Ma (2018, GSA timescale v. 5.0). The Precambrian takes up about 88.12% of the entire history of the Earth (based off Figures 1 and 2). The largest division of geologic time is the "eon" via the 2005 version of the North American Stratigraphic Code (NASC). The NASC is the only guideline for formalizing geologic time. Its "non material" based standards are nowhere near as stringent as they are for "material based" units such as lithostratigraphic divisions.

The Precambrian eons recognized herein, from oldest to youngest are; the Hadean (formalized herein) which spanned from 4,560-3,810Ma (covering 750 million years, 16.45% of Earth's total history, 18.66% of the Precambrian); the Archean (redefined herein) which spanned from 3,810-2,487Ma (covering 1,323 million years, 29.01% of Earth's total history, 32.92% of the Precambrian); and the Proterozoic (redefined herein) which spanned from 2,487-541Ma (covering 1,946 million years, 42.67% of Earth's total history, 48.42% of the Precambrian). Some era integrity was attempted. I didn't want to toss the timescale into total chaos, plus it wasn't necessary to revamp the timeline at that scale.

I attempted not to move any of the era boundaries too far from their 2018 GSA timescale position. The only exception to this is the Hadean-Archean boundary. I moved it back up from 4,000Ma 2018, GSA timescale v. 5.0 up to 3,810Ma, which is closer to the 2009 GSA time scale position of 3,850Ma. I did this not because it's my best guess of when the heavy bombardment ended, I did it based off of the Isua Greenstone belt which is presently the first known occurrence of the oldest supracrustal rocks, which are defined off the North Atlantic Craton of Greenland (see Figure 2). I am dominantly trying to frame time based off well dated events, not estimates. Even if older supracrustal rocks are found in the future, somewhere else in the world, the border is herein based off the Greenland rocks of the North Atlantic Craton.

That is the point of these redefinitions or new names. It was to define the chronological borders based off well defined dates within a specific craton or an area within the craton. It wasn't to set them at events that are likely to fluctuate wildly. It's tempting to use monumental events like; the end of the heavy bombardment, the first appearance of a craton, or the beginning of Plate Tectonics. However, the definitions of these events are not totally set in stone, nor are they firmly locked into place, nor do they have a firm age they can be locked into. There's no impact crater I can point to and say, "that one was the last one of the heavy bombardment". Even if we presently accepted a date of 3,800Ma for the end of the Heavy bombardment, we could find a big crater tomorrow that dates at 3,650Ma and argue that was the end of it. I wanted to avoid those type of things by using targeted quantitative definitions over vague qualitative definitions. I do use a couple where unavoidable or significant. Like the 4,510Ma date in between the Hadean's Chatonian and Lunarian Periods (Figure 2). But that is based off a single impact that formed the moon and isn't likely to wildly change.

Time slices in the Precambrian, will by nature, typically be a lot longer than their Phanerozoic counterparts. This is due to the fact that the further you go back in time, the less complete the rock record generally is. The longest era of the Phanerozoic is the Paleozoic, which is ~289 million years in length (2018, GSA timescale v. 5.0). In contrast the longest era in the Precambrian is the Mesoproterozoic with a 790 million year duration (see Figure 2). The 2018 GSA timescale takes time units down to as small as the "age". Table 1 shows the formally recognized chronologic time divisions.

<u>Table 1</u>: Breakdown of chronologic units

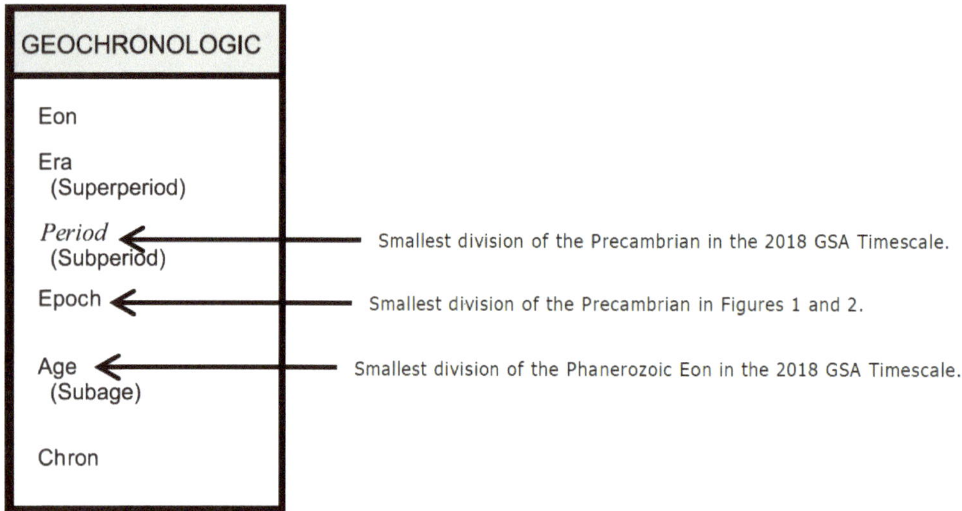

Adapted from the 2005 North American Stratigraphic code. Italicized = fundamental unit.

In the Precambrian, there is no need to go smaller than the Epoch. Even herein the smallest division of time in the Hadean and Archean is the "period". Further Epoch divisions of Figure 2, may be possible in the future. I did not feel it was necessary at this time.

Every reasonable attempt has been made to either redefine and existing name or give a unit of time a new name if the redefinition involved an effective dissolution of a timeframe as defined by the GSA (2018). The only exception to this is the Epochs in the Paleoproterozoic. They have not yet been given names, although their future division is warranted based off the prominent glacial cycles during that time. They have only been given standard geologic map designations (FGDC, 2006). The reason being, was to avoid repeating names. Although there is nothing technically wrong with using a lithostratigraphic term (e.g. "Huronian Supergroup") for a chronostratigraphic term (e.g. a hypothetical "Huronian Epoch"), it can still lead to confusion, and should be avoided. So until suitable names can be found, I used map symbols for the epochs of the Paleoproterozoic (Figure 1). For example, The Eoarchean of the 2018 GSA timescale has been abandoned (Figure 2). With the youngest date of the Hadean being moved up from 4,000 to 3,810 million years, this effectively split the Eoarchean. Instead of just shrinking it, I eliminated it and shifted the Paleoarchean.

Mesoproterozoic red clastic interflow sandstone overlain by gray volcanics along Minnesota Route 61. Photo taken by Steven Baumann on May 10, 2019.

As we gain knowledge of the Precambrian rocks, a precise timescale is not only practical, but inevitable. We as humans like to give things names so we can talk about them. It's a lot easier to talk about the Paleoproterozoic, Mesoproterozoic, and Neoproterozoic, than it is the X, Y, and Z Eras (the map symbols for these eras; FGDC, 2006). The subdivisions of the Precambrian need defined and set boundaries. That is what is attempted herein.

METHOD

The defining of boundaries is a nomenclature and standardization issue much more than it is a methodological one. However, it needs to be based off of something so there needs to be a method, or reason, as to the redefinitions. The use of material referents will play a part, as defined by the NASC (2005). In order to define the eons, eras, periods, and epochs in the Precambrian; we need to set a universal standard based off of something more than just arbitrary evenly spaced boundaries. As I have already stated, using major global events with poorly constrained ages, and isn't practical. Although I am redefining the ages of things based off of geochronology (non material based), it is tied to chronostratigraphy, which is in part material based (NASC, 2005). Using material referents for geochronologic units, is appropriate.

We could do what has traditionally been done and assign arbitrary boundaries, which has been traditionally done and is one of Bleeker's (2004) criticisms. Bleeker as well as Kranendonk (2012) were highly influential in developing this concept. However, I disagree strongly with Bleeker's (2004) claim that using U-Pb ages are "naïve". He suggests using Pb-Pb dating, which I agree can also be used. Yes, they can have margins of error up to $\pm$10 million years (although nowadays often much smaller); but when dealing with the Precambrian, that's negligible. What you do in that case, as Bleeker himself eluded to (2004), you pick a stratigraphic boundary within the margin of error, if possible. This works really well with the Proterozoic, but breaks down a bit in the Hadean and Archean, as we have a much smaller sedimentary rock assemblage to deal with. In that case, you just set the boundary at the value excluding the margin of error. Later, other workers can redate the rock unit using better methods to lock down a firmer boundary. Adjusting a boundary in the Precambrian by a few million or even 10 million years, isn't that big of a deal.

I do agree with Bleeker that we need a natural, or more stratigraphic (material) basis for even selecting these dates. What is meant by that? It means we should use well dated individual and very localized lithostratigraphic units to define the boundaries and not more regional events like the "appearance of the first craton" That has the potential to be either significantly moved or discarded altogether if an older one were found. You can use a lithostratigraphic unit that has been well dated on your craton, but not the craton itself. It doesn't matter where in the world we pick these well bracketed lithostratigraphic units, as long as they are well constrained. Thick rock units with a dominantly sedimentary protolith with internal conformities (like the Huronian Supergroup) are ideal for this (Baumann, 2019; and references therein). That isn't always practical, so it becomes necessary to use things like well dated plutons especially in the Hadean and Archean (Figure 2).

When naming new divisions of time, just slapping on "-ian" or "-an" to a lithostratigraphic unit is not only impractical, but violates article 77 of the most recent version (NASC, 2005). Using lithostratigraphic units for chronological divisions causes confusion and should be avoided. I also recommend new names and not using other things, such as the name of a supercontinent or a geographic region (like a craton or continent). The reason for not using them is the same as for lithostratigraphic units, to avoid duplicate names and to eliminate nomenclature confusion. Also, you have to use local geographic names for lithostratigraphic units (NASC, 2005); by using local places for geochronologic names, you are removing them from the potential "pool of names". Articles 80-82 of the NASC (2005) define the standard for naming geochronologic units. Using Greek or Latin words is not mandatory, neither is using made up names against the standard. The only thing that would violate the code is if it has a named chronostratigraphic unit and I don't use that, because something like a "system (chronostratigraphic) = a period (geochronic)". This isn't really an issue in the Precambrian, since chronostratigraphic units haven't been locked down. No "stratotype or "reference section" is required for geochronologic units.

The use of "transition zones" (Bleeker, 2004), should be avoided. Gray area transition zones are fine for lithostratigraphic units (often referred to as gradational zones; NASC, 2005), but they make no sense in a geochronologic classification.

Time moves forward in a well defined "tick-tock" pace. There are no "gray areas". There might be gaps or gradations in the rock record, but that has zero effect on the progression of time moving forward. They aren't necessarily based off of "material referents" (NASC Article 65, 2005) a.k.a. rock units, so there is no need for transition zones. If a "transition zone" is suspected, it is better to either make it is own geochronologic unit, or set the boundary at the top or bottom.

I suggest the following become the standard for picking geochronologic boundaries within the Precambrian:

1) Boundaries should be based of well dated events; e.g. well constrained rocks whose protoliths are sedimentary packages, individual impacts, individual plutons, or lava flows. The boundaries should be put at the beginning or end dates of deposition or the date of emplacement, not metamorphic events.
2) Well constrained and firmly dated individual large impact events can be used as boundaries. Ones with reconstructed craters >100km (>~62 miles) in outer rim diameter.
3) An arbitrary value in between well constrained unconformities with <40 million years of missing time can be used. Although not necessary, picking the middle point is the best option.
4) Local rock units in some way need to be used. Large scale global unconstrained events (e.g. the heavy bombardment, the first appearance of cratons, the assemblage of supercontinents, etc.) are not to be used. However, specific well constrained geologic units within them, can be used.
5) Only units with well constrained dates can be used. It doesn't matter what radiometric dating method was used. As long as the margin of error is relatively small and it isn't in dispute, it can be used.
6) New names for time divisions should only be used if that time unit is subdivided. An undivided time unit should not be given another name at one level down (Table 2).

Kranendonk (2012), was heavily relied on as well as the references therein. Figure 16.34 on page 361, was often used for guidance and names.

Table 2: Kranendonk's 2012 timescale (modified)

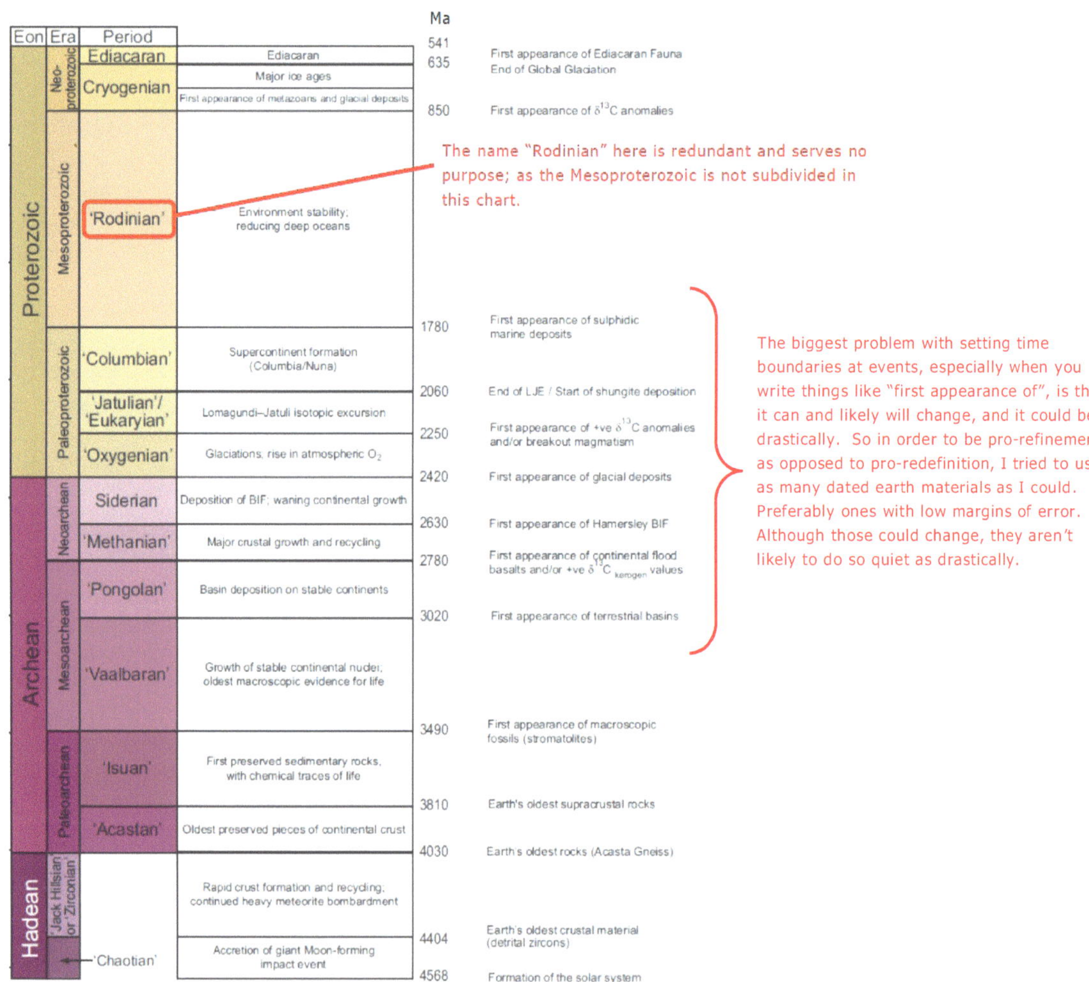

Kranendonk's "**FIGURE 16.34**: Proposed scheme for a fully revised Precambrian timescale."

The Hadean is herein formalized as the first eon in Earth's history and is bracketed from the formation of the Earth at about 4,560Ma (Kranendonk, 2012); to 3,810Ma, which is the oldest well preserved supracrustal rocks in the North Atlantic Craton of Greenland (Kranendonk, 2012). Table 3 shows the breakdown of the spans of time in the Hadean. Figure 2, shows the marker events.

The Hadean is herein divided into the Paleohadean, Mesohadean, and Neohadean Eras. The Paleohadean is from the formation of the Earth to the oldest known terrestrial zircon in the Jack Hills area at 4,463Ma (Kranendonk, 2012). The Paleohadean has two herein redefied periods, the older Chatonian and the younger Lunarian, they are separated by the impact that formed the moon at 4,510Ma (Kranendonk, 2012).

The Mesoarchean is from the end of the Paleohadean Era to the graphite in the Jack Hills area in Australia, which is about 4,100±10Ma. This could be the oldest organically derived carbon in the rock record (Bell et al., 2015). The Mesohadean is not subdivided into periods at this time.

The Neohadean lower boundary is at the Mesoarchean. Its upper boundary is at 3,810Ma, which is based off the presently oldest known preserved supracrustal rocks. This isn't what actually defines the upper boundary, it's just an event associated with it. The protolith age of the tonalitic gneiss in the Isua Greenstone Belt of Greenland is the upper age at 3,810Ma (Crowley, 2003). The Neoarchean is further subdivided into the older Zirconian (Kranendonk, 2012) and younger Acastan (Kranendonk, 2012) Periods. The division of the two is the oldest protolith age of the Acasta Gneiss in the Slave Craton (Kranendonk, 2012).

Table 3: Timespans of the Hadean Eon

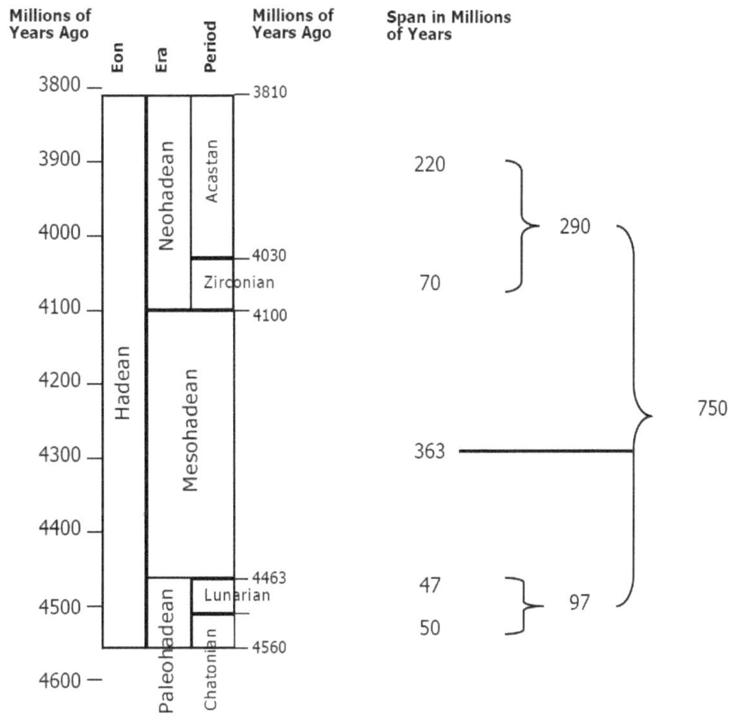

The lower boundary of the Archean Eon (herein redefined) is the upper boundary of the Hadean Eon. The upper boundary of the Archean herein is presently set at 2,487Ma, and is based off the oldest volcanics of the Huronian Supergroup in Ontario (Baumann, 2019; and references therein).

The Archean is divided into three eras, the oldest to youngest are; the Paleoarchean, Mesoarchean, and Neoarchean. The Eoarchean is herein abandoned as the Hadean was lengthened at the Archean's expense (Figure 2).

The Paleoarchean (herein redefined) is from the top of the Hadean to 3,490Ma which is the first appearance of stromatolites in the metasediments of the Pilbara Craton (Kranendonk, 2012). It is not subdivided into periods or epochs at this time.

The Mesoarchean is herein redefined as the longest Era within the Precambrian, covering a span of 750 million years (Figure 2). Its lower boundary is the top of the Paleoarchean. Its upper boundary is presently at 2,740Ma, and is based off the oldest tholeiitic rocks of the South Central Slave Superterrane in northwestern Canada (Helmstaedt and Pehrsson, 2012). It is subdivided into three periods. From oldest to youngest they are the Stromatolanian (named herein after stromatolites) ranging from 3,490 to 3,236Ma. The upper boundary of 3,236Ma is based off the emplacement of the Nelshoogte Tonalite-trondhjemite-granodiorite (TTG) pluton in the Kaapvaal Craton (Anhaeusser, 2006). The Kaapian, named herein after the Kaap Valley pluton in South Africa, began at the end of Stromatolanian and ended with the oldest deposits of the DeGrey Supergroup in the Pilbara Craton of South Africa. This sets it at 3,020Ma (Kranendonk, 2012). The Kaapian was followed immediately by the youngest of the Mesoarchean periods, the Cratonian, named herein as it was the first era where cratonic basins form. The Cratonian ranged from the end of the Kaapian to the end of the Mesoarchean at 2,740Ma.

The youngest era of the Archean Eon is the Neoarchean (redefined herein). Its lower boundary is the upper boundary of the Mesoarchean (2,740Ma). Its upper boundary is the upper boundary of the Archean Eon (2,487Ma). It is divided into two periods. The oldest is the Methanian revised from Kranendonk, 2012. The youngest is named herein as the Nneuonian, "neu" is the German word for new. The boundary between the periods is locked in at the youngest orogenic granitic rock of the Hackett River Terrane in the Slave Craton; which is currently at 2,580Ma (Helmstaedt, 2009).

Archean migmatite and gneiss along Trans-Canada 17 in Ontario, Canada.
Photo taken by Steven Baumann on May 9, 2019.

The Proterozoic is the longest spanned Eon in the geologic timescale. Its beginning is at the end of the Archean Eon and its ending is at the beginning of the Phanerozoic Eon (Figure 1). That has not changed herein. What has changed is the end of the Archean. Herein the Proterozoic is bracketed at 2,487 and 541Ma, a span of 1,946 million years. This publication only deals with the two oldest Eras of the Proterozoic, the Paleoproterozoic and the Mesoproterozoic. The Neoproterozoic is not included, as it is by far the most dynamic of the Precambrian eras on the GSA timescale. Unlike the Hadean and Archean, most of the Paleoproterozoic is divided into Epochs.

Paleoproterozoic

The Beginning pf the Paleoproterozoic is defined almost entirely off the Huronian Supergroup in Ontario. Its lower boundary is the upper boundary of the Archean at 2,478Ma; which is the oldest of the Huronian rift volcanics. Its upper boundary is to completion of the "continental margin deposits". That seems a bit vague, but it isn't. A lot of stuff was going on at this time. The Penokean Orogeny in Wisconsin and the Upper Peninsula of Michigan was complete, and the deposition of the forearc deposits, known as the "continental assemblage" had been all but completed. This isn't a strong enough date. Shortly after this, the post-Penokean granites like the Amberg, dated at $1,754\pm11$ (Holm, et al., 2012), were being deposited. This still isn't tight enough and that magmatic event lasted from ~1,759 to ~1,750Ma. With the exception of a rhyolite, all five related igneous rocks have large margins of error (Holm, et al., 2012). It's tempting to use that, bet there is a better set of slightly older plutons to use that lie in between the two. The batholiths of east central Minnesota can be used, and specifically the Richmond Granite. There are 23 dated plutons, with many of them having margins of error of <10 million years. All were deposited within a relatively short window of time at about 24 million years, from about ~1,796 to ~1,772Ma (Holm, et al., 2012). The Richmond Granite has been dated at $1,772\pm3$Ma. This makes it a good marker because of its low margin of error and it is suspected as being the youngest pluton of the east-central Minnesota batholiths.

The Paleoproterozoic is divided into five periods. From oldest to youngest they are the Siderian (redefined herein), the Oxygenian (redefined from Kranendonk, 2012), Jatulian (redefined from Kranendonk, 2012), Orosirian (redefined herein) and the Marginian (defined herein).

The Siderian appears on the GSA 2018 timescale and has been restricted here. The lower boundary is the base of the Proterozoic and the upper boundary is within the unconformity separating the Cobalt and Quirke Lake Groups of the Huronian Supergroup (Baumann, 2019; and references therein). The Siderian (Xs) is subdivided into three unnamed epochs (Figure 1) and they are all separated by unconformities within the Huronian Supergroup (Baumann, 2019; and references therein). They are given map designations of Xs_1, Xs_2, and Xs_3 (respectively); until suitable names are found.

The Oxygenian (Kranendonk, 2012) is also restricted to the Cobalt Group of the Huronian Supergroup. During this time we had free oxygen appearing in the atmosphere, hence the name. However, that couldn't be used to define it. So its lower boundary is the top of the Siderian and its upper boundary is the oldest Nipissing intrusion which presently is dated at 2,219.4+3.6/-3.5Ma (Corfu and Andrews, 1986), so 2,219Ma is adopted. It is not divided into epochs.

The Jatulian has its lower boundary at the top of the Oxygenian. Its upper boundary is the start of the Glenberg Orogeny, which is marked by the maximum age of the metapelite included in diatexite at $2,083\pm7$Ma (Johnson, et al., 2010). The Jatulian is not divided into epochs.

The Orosirian has herein been redefined. Its lower boundary is the top of the Jatulian. Its upper boundary is set at the Sudbury impact event at $1,850\pm1$Ma (Krogh, et al., 1984). It is subdivided into four epochs, given map designations until suitable names can be assigned. From oldest to youngest they are Xo_1, Xo_2, Xo_3, and Xo_4 (Figure 1). The boundary between Xo_1 and Xo_2 is set at the Vredefort impact event which has been dated at $2,023\pm4$Ma (Kamo, et al., 1996). The boundary between Xo_2 and Xo_3 is the youngest depositional age of the Quartpot Pelite of the Cama Hills in western Australia (Spaggiari and Kirkland, 2014). It was deposited in a forearc basin at the end of the Glenberg Orogeny. This is similar to the slightly younger continental margin deposits (e.g. Michigamme Formation) of the Penokean Orogeny. The boundary between Xo_3 and Xo_4 is set at the beginning of the Penokean Orogeny at 1,895Ma (Baumann, 2019a; and references therein). Why am I not using a dated igneous event? The oldest dated igneous rock is from a rhyolitic part of the Quinnesec Diorite, which has been dated at $1,866\pm39$Ma (Peterman, et al., 1985). That is a huge margin of error. I could have used the +39 and set it at 1,905Ma, but that isn't necessary in this case, and would arguably not be following my own standard. Based on the dated volcanics and plutons of the Penokean Orogeny, about a dozen in total (Baumann, 2019a), we have a reliable beginning of the Penokean Orogeny at 1,895Ma.

The youngest of the Paleoproterozoic periods is herein named the Marginian Period because of the thick continental margin sediments laid down at this time (e.g. the Michigamme, Tyler, and Rove Formations) in the Lake Superior area. The Sudbury impact marks its lower boundary (1,850Ma). The end of the Paleoproterozoic (1,772Ma) marks its upper boundary. It is not divided into epochs.

<u>Mesoproterozoic</u>

The moving of the Statherian Period brings us into the Mesoproterozoic Era. Its beginning is the end of the Paleoproterozoic Era. Its end is the beginning of the Neoproterozoic Era, which herein is set at 980Ma. 980Ma is somewhat tentative at present but is based off Malone's et al. (2016) interpretation of the age of the Jacobsville Group as defined by Baumann, et al. (2016). This will most likely need to be refined in the future. It is based off the youngest possible age for the Freda Sandstone in the Upper Peninsula of Michigan and Wisconsin. Although, not presently locked down, it hopefully will be within the next couple of years as I complete my work on the Freda.

The Mesoproterozoic is divided into four periods. From oldest to youngest they are the Statherian, Calymmian, Ectasian, and Stenian. All are redefined, from the GSA 2018 timescale, herein.

The Statherian Period as defined in the 2018 GSA timescale, has been removed from the Paleoproterozoic, redefined herein, and moved to the Mesoproterozoic Era. This was done for a couple of reasons. First being that from this point forward North America, a.ka. Laurentia, would grow almost entirely to the south and the east through accretionary terranes, until the Phanerozoic (Whitmeyer and Karlstrom, 2007). Second, the geologic movements of the continents during this time was leading to the formation of the first confirmed supercontinent where >85% of cratons would be assembled into one "megacontinent" called Rodinia (Li, et al., 2008). The third reason is because of a period of time that geologists refer to as the "boring billion" or "barren billion". This period of time was from about ~1,850 to ~850Ma (Roberts, 2013, and references therein). The development of life seems to have been profoundly effected by this relatively "calm" part of Earth's geologic past (Mukherjee, et al., 2018). The barren billion isn't actually one billion years, but its close. By sticking the Statherian into the Mesoproterozoic, I am acknowledging the barren billion as a significant span of time in Earth's geologic past, but I am not using it as a standard for time divisions. Its lower boundary at the top of the Paleoproterozoic. Its upper boundary is set at the emplacement of the Monte Largo Pluton in New Mexico which has been dated at $1,656\pm10$Ma (Bauer, et al., 1993). It is not divided into epochs at this time.

The Calymmian Period is herein redefined. Its beginning is at the end of the Statherian Period. Its top is set at the end of deposition for the youngest preserved sedimentary unit of the Missoula Group, the Garnet Range Formation, in Montana; specifically within Glacier National Park, which is currently set at 1,300Ma (Baumann, in progress). It is separated into two epochs, an older Basinian epoch. The Basinian is named herein because during this time the world began to see the formation of massive intracratonic basins that subsided far faster than any Phanerozoic basins; such as the basins where the Belt Supergroup (Montana), Sibley Group (Ontario), and Pingicula Group (Yukon). Some were associated with rifting, some were not. The Ontanan (named herein for the Canadian Province of Ontario and the American State of Montana) is the younger of the two epochs. The boundary between the two is set at the beginning of sedimentary deposition of the Lower Belt Group of the Belt Supergroup in Montana, which is at 1,470Ma (Baumann, in progress).

Above the Calymmian is the Ectasian. The Ectasian is herein redefined. Its lower boundary is the top of the Calymmian and its upper boundary is the oldest dated midcontinent rift (MCR) volcanic unit, which has been measured at 1,136Ma (Zartman, et al., 2011). The Ectasian is comprised of the older Arbucklian Epoch. Which is named herein for the Arbuckle Mountains in Oklahoma that contain the Troy Granite, which has been dated at $1,399\pm95$Ma (Thomas, et al., 2012). The younger Gardarian Epoch. Which is named herein for the Gardar area (now called Igaliku) of Greenland where a failed rift sits. The boundary between the two is the youngest of the Timmiarmiit Dikes in Gardar Greenland, which have been dated to $1,268\pm4$Ma (Bartels, et al., 2016).

The end of the Mesoproterozoic is marked by the Stenian Period (redefined herein). It begins at the end of the Ectasian and ends where the Neoproterozoic begins, which is presently set at 980Ma. It is herein separated into two epochs, the older Gogebician and the younger Bayfieldian. The Gogebician is named after Gogebic County, Michigan, where over a billion and a half years of geologic history is exposed to include the volcanics and the clastic sediments of the MCR. The younger is the Bayfieldian Epoch, named after Bayfieldian County, WI where the Precambrian clastic sediments on top of the MCR are well exposed. The "Bayfield" was a group until it was redefined along with the Jacobsville and discarded (Baumann, et al., 2016). Once a name is discarded it can be reused for other units (NASC, 2005). The two epochs are separated by the end of dominantly volcanic deposits of the MCR. This is also the boundary between the Beaver Bay Complex, dated at $1,091.61\pm0.14$Ma and the dated andesite of the Lake Shore Traps (within the Copper Harbor Formation) dated at $1,085.57\pm0.25$Ma (Fairchild, et al., 2017). The mean average of those two dates, excluding margins of error) is 1,088.59Ma, which is about 1,089Ma.

The graphical representations of all the changes herein are in Figures 1 and 2.

Neoproterozoic Jacobsville Group in contact with a black igneous intrusion on Presque Isle, Marquette, Michigan. Photo taken by Steven Baumann on May 4, 2019.

FIGURE 1: Proposed Adjustments to the Precambrian Proterozoic Timescale (Based on the 2018 5.0 GSA Timescale)

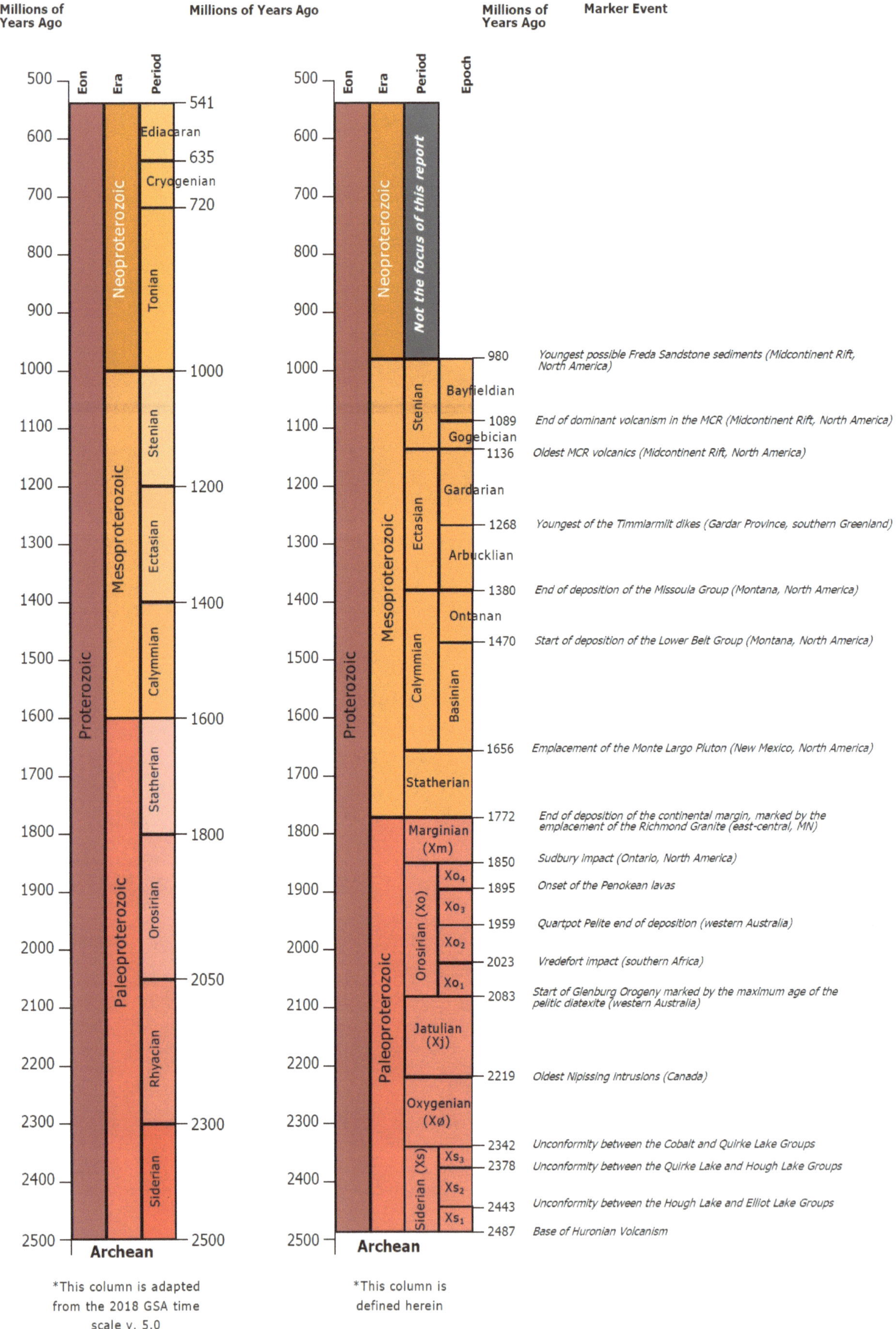

Millions of Years Ago
Millions of Years Ago
Millions of Years Ago
Marker Event

Eon
Era
Period

500
500
541
Ediacaran
600
635
Cryogenian
700
720
800
Neoproterozoic
Tonian
900
1000
1000
Stenian
1100
Mesoproterozoic
1200
1200
Ectasian
1300
1400
1400
Calymmian
1500
Proterozoic
1600
1600
Statherian
1700
1800
1800
Paleoproterozoic
Orosirian
1900
2000
2050
2100
Rhyacian
2200
2300
2300
Siderian
2400
2500
2500
Archean

*This column is adapted from the 2018 GSA time scale v. 5.0

Eon
Era
Period
Epoch

500
500
600
Neoproterozoic
Not the focus of this report
700
800
900
1000
980   Youngest possible Freda Sandstone sediments (Midcontinent Rift, North America)
Stenian   Bayfieldian
1100
1089   End of dominant volcanism in the MCR (Midcontinent Rift, North America)
Gogebician
1136   Oldest MCR volcanics (Midcontinent Rift, North America)
1200
Gardarian
Ectasian
1268   Youngest of the Timmiarmiit dikes (Gardar Province, southern Greenland)
1300
Mesoproterozoic
Arbucklian
1380   End of deposition of the Missoula Group (Montana, North America)
1400
Ontanan
1470   Start of deposition of the Lower Belt Group (Montana, North America)
1500
Calymmian
Basinian
1600
Proterozoic
1656   Emplacement of the Monte Largo Pluton (New Mexico, North America)
1700
Statherian
1772   End of deposition of the continental margin, marked by the emplacement of the Richmond Granite (east-central, MN)
1800
Marginian (Xm)
1850   Sudbury impact (Ontario, North America)
Xo4   1895   Onset of the Penokean lavas
1900
Xo3   1959   Quartpot Pelite end of deposition (western Australia)
Orosirian (Xo)
Xo2   2023   Vredefort impact (southern Africa)
2000
Xo1   2083   Start of Glenburg Orogeny marked by the maximum age of the pelitic diatexite (western Australia)
2100
Paleoproterozoic
Jatulian (Xj)
2200
2219   Oldest Nipissing intrusions (Canada)
Oxygenian (Xø)
2300
2342   Unconformity between the Cobalt and Quirke Lake Groups
Xs3   2378   Unconformity between the Quirke Lake and Hough Lake Groups
2400
Siderian (Xs)   Xs2
2443   Unconformity between the Hough Lake and Elliot Lake Groups
Xs1   2487   Base of Huronian Volcanism
2500
2500
Archean

*This column is defined herein

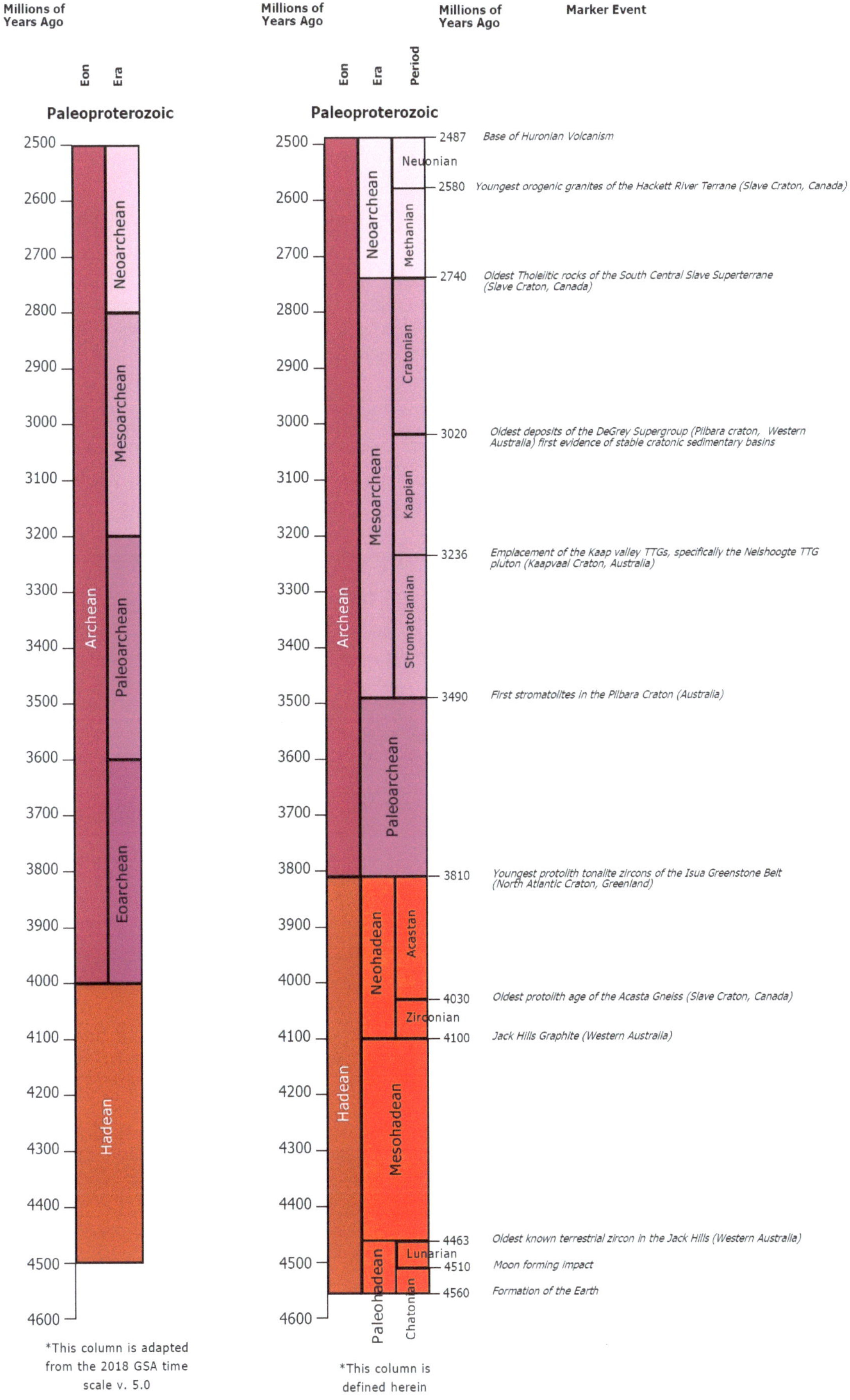

FIGURE 2: Proposed Adjustments to the Precambrian Archean-Hadean Timescale (Based on the 2018 5.0 GSA Timescale)

Millions of Years Ago
Millions of Years Ago
Millions of Years Ago
Marker Event

Eon
Era

Eon
Era
Period

Paleoproterozoic
Paleoproterozoic

2500
2600
2700
2800
2900
3000
3100
3200
3300
3400
3500
3600
3700
3800
3900
4000
4100
4200
4300
4400
4500
4600

Archean
Neoarchean
Mesoarchean
Paleoarchean
Eoarchean

Hadean

2500
2600
2700
2800
2900
3000
3100
3200
3300
3400
3500
3600
3700
3800
3900
4000
4100
4200
4300
4400
4500
4600

Archean
Neoarchean
Mesoarchean
Paleoarchean

Hadean
Neohadean
Mesohadean
Paleohadean

Neuonian
Methanian
Cratonian
Kaapian
Stromatolanian
Acastan
Zirconian
Lunarian
Chatonian

2487   Base of Huronian Volcanism
2580   Youngest orogenic granites of the Hackett River Terrane (Slave Craton, Canada)
2740   Oldest Tholeiitic rocks of the South Central Slave Superterrane (Slave Craton, Canada)
3020   Oldest deposits of the DeGrey Supergroup (Pilbara craton, Western Australia) first evidence of stable cratonic sedimentary basins
3236   Emplacement of the Kaap valley TTGs, specifically the Nelshoogte TTG pluton (Kaapvaal Craton, Australia)
3490   First stromatolites in the Pilbara Craton (Australia)
3810   Youngest protolith tonalite zircons of the Isua Greenstone Belt (North Atlantic Craton, Greenland)
4030   Oldest protolith age of the Acasta Gneiss (Slave Craton, Canada)
4100   Jack Hills Graphite (Western Australia)
4463   Oldest known terrestrial zircon in the Jack Hills (Western Australia)
4510   Moon forming impact
4560   Formation of the Earth

*This column is adapted from the 2018 GSA time scale v. 5.0

*This column is defined herein

<u>REFERENCES</u>

Anhaeusser, C.R., 2006. *A reevaluation of Archean intracratonic terrane boundaries on the Kaapvaal Craton, South Africa. Collisional suture zones?*, Special Paper of the Geological Society of America, DOI:10.1130/2006.2405(11)

Baumann, S.D.J., Cory, A.B., and Dylka, S.K., 2016: Lithostratigraphic interpretation and redefinition of the sedimentary clastic assemblage (Bayfield Group and Jacobsville Sandstone) in Michigan and Wisconsin, USA and Ontario, Canada, Journal of Stratigraphy, v.13, no.3

Baumann, S.D.J. (compiled by), 2019: *Stratigraphic column and ages of the Huronian Supergroup and Chocolay Group*, Midwest Institute of Geosciences and Engineering, Publication G-082017-1C

Baumann, S.D.J., 2019a: *Geochronology of the main deposits of the Penokean Orogeny of northeastern Wisconsin*, Midwest Institute of Geosciences and Engineering, publication G-112019-2A

Bartels, A., Nilsson, M.K.M, Klausen, M.B., and Soderlund, U., 2016: Mesoproterozoic dykes in the Timmiarmiit area, Southeast Greenland: evidence for a continuous Gardar dyke swarm across Greenland's North Atlantic Craton, Journal of the Geological Society of Sweden, GFF, v.138, i1, p.255-275, DOI:10.1080/11035897.2015.1125386

Bauer, P.W., Karlstrom, K.E., Bowing, S.A., Smith, A.G., and Goodwin, L.B., 1993: Proterozoic plutonism and regional deformation-new constraints from the Southern Manzano Mountains, central New Mexico, New Mexico Geology, v.15, no.3

Bell, E.A., Boehnke, P., Harrison, T.M., and Mao, W.L., 2015: *Potentially biogenic carbon preserved in a 4.1 billion-year-old zircon*, Proceedings of the National Academy of Sciences (PNAS) of the U.S.A., v.112, no.47, DOI:10.1073/pnas.1517557112

Bleeker, W., 2004: *Towards a 'natural' time scale for a Precambrian-A proposal*, v.37, p.219-222, Oslo. ISSSN 0024-1164

Bleeker, W., Ketchum, J., Davis, B., Sircombe, K., Stern, R., Waldron, J., 2004a: *The Slave Craton from on top: the crustal view*

Corfu, F. and Andrews, A.J., 1986: *A U-Pb age for mineralized Nipissing diabase, Gowganda, Ontario*, Canadian Journal of Earth Sciences, 23(1), p.107-109, DOI:10.1139/e86-011

Crowley, J.L., 2003: *U-Pb geochronology of the 3810-3630Ma granitoid rocks south of the Isua greenstone belt, southern West Greenland*, Journal of Precambrian Research, v.126, i3-4, p.235-257, DOI:10.1016/S0301-9268(03)00097-4

Fairchild, L.M., Swanson-Hysell, N.L., Ramezani, J., Sprain, C.J., and Bowring, S.A., 2017: *The end of midcontinent rift magmatism and the paleogeography of Laurentia*, Lithosphere, v.9, no.1, p.117-133, DOI:10.1130/L580.1

Helmstaedt, H., 2009: *Crust-mantle coupling revisited: the Archean Slave Craton, NWT, Canada*, Lithos, DOI:10.1016/j.lithos.2009.04.046

Helmstaedt, H.H., and Pehrsson, S.J., 2012: *Geology and tectonic evolution of the Slave Province-a post lithoprobe perspective*, Geological Association of Canada, Special Paper 49, Chapter 7, p.379-466

Holm, D.K., Van Schmus, W.R., MacNeill, L.C., Boerboom, T.J., Schweitzer, D., and Schneider, D., 2012: *U-Pb zircon geochronology of Paleoproterozoic plutons from the northern midcontinent, USA: Evidence for subduction flip and continued convergence after geon 18 Penokean orogenesis*, Geological Society of America bulletin 2005;117, no.3-4, p.259-279, DOI:10.1130/B25395.1

Johnson, S.P., Sheppard, S., Rasussen, B., Wingate, M.T.D., Kirkland, C.L., Muhling, J.R., Fletcher, I.R., and Belousova, E., 2010: *The Glenburgh Orogeny as a record of Paleoproterozoic continent-continent collision*, Government of Western Australia, Department of Mines and Petroleum, Record 2010/5

Kamo, S.L., Reimold, W.U., Krogh, T.E., and Colliston, W.P., 1996: *A 2.023Ga age for the Vredefort impact and a first report of shock metamorphosed zircons in pseudotachylitic breccias and granophyre*, Earth and Planetary Science letters, v.144, i.3-4, p.369-387, DOI:10.1016/S0012-821X(96)00180-X

Krogh, T.E., Davis, D.W., and Corfu, F., 1984: *Precise U-Pb zircon and baddeleyite ages in the Sudbury area*, The Geology and ore deposits of the Sudbury structure (Special volume / Ontario Geological Survey), Chapter 20, p.431-446

REFERENCES (Continued)

Li., Z.X., Bogdanova, S.V., Collins, A.S., Davidson, A., De Waele, B., Ernst, R.E., Fitzsimons, I.C.W., Fuck, R.A., Gladkochub, D.P. Jacons, J., Karlstrom, K.E., Lu, S., Natapov, L.M., Pease, V., Pisarevsky, S.A., Thrane, K., and Vernikovsky, V., 2008: *Assembly, configuration, and break-up history of Rodinia: A synthesis*, Journal of Precambrian Research, v.160, p.179-210

Malone, D.H., Stein, C.A., Craddock, J.P., Kley, J., Stein, S., and Malone, J.E., 2016: *Maximum depositional age of the Neoproterozoic Jacobsville Sandstone, Michigan: Implications for the evolution of the Midcontinent Rift*, Geosphere, v.12, no.4, 6 giures, 1 table, DOI:10.1130/GES01302.1

Mukherjee, I., Large, R.R., Corkrey, R., and Danyushevsky, L.V., 2018: *The boring billion, a slingshot for complex life on Earth*, Scientific Reports, 8:4432, DOI:10.1038/s41598-018-22695-x

Peterman, Z.E., Sims, P.K., Zartman, R.E., and Schulz K.J., 1985: *Middle Proterozoic uplift events in the Dundar dome of northeastern Wisconsin, USA*, Contributions to Mineralogy and Petrology, v.91, p.138-150

Roberts, M.W., 2013: *The boring billion? -Lid tectonics, continental growth and environmental change associated with the Columbia supercontinent*, Geoscience frontiers, v.4, p.681-691

Spaggiari, C.V., Kirkland, C.L., Smithies, R.H., and Wingate, M.T.D., 2014: *Tectonic Links between Proterozoic sedimentary cycles, basin formation and magmatism in the Albany-Fraser Orogen, Western Australia*, Government of Western Australia, Department of Mines and Petroleum, Report 133

Thomas, W.A., Tucker, R.D., Astini, R.A., and Denison, R.E., 2012: Ages of pre-rift basement and synrift rocks along the conjugate rift and transform margins of the Argentine Precordillera and Laurentia, Geosphere, Making the Southern Margin of Laurentia themed issue, v.8, no.6, p.1366-1383, DOI:10.1130/GES00800.1

Various authors, 2005: *North American stratigraphic code*, North American Commission on Stratigraphic Nomenclature, American Association of Petroleum Geologists (AAPG) bulletin, v.89, no.11, p.1547-1591, DOI:10.1306/07050504129

Various authors, 2006: *Federal Geographic Data Committee (FGDC) digital cartographic standard for geologic map symbolization*, Reston, VA, FGDC document number FGDC-STD-013-2006, 290p., 2 plates

Various authors, 2018: *Geologic Society of America timescale*, V.5.0

Van Kranendonk, M.J., 2012: A chronostratigraphic division of the Precambrian, The geologic time scale 2012, Chapter 16, v.1, p.299-392, ISBN: 978-0-44-459390-0

Whitmeyer, S.J., and Karlstrom, K.E., 2007: Tectonic model for Proterozoic growth of North America, Geosphere, v.3, no.4, p.220-259, DOI: 10.1130/GES00055

Zartman, R.E., Nicholson, S.W., Cannon, W.F., and Morey, G.B., 2011: *U-Th-Pb zircon ages of some Keweenawan Supergroup rocks from the south shore of Lake Superior*, Canadian Journal of Earth Sciences, v.34, no.4, p.549-561

Preliminary Geologic Map of Little Presque Isle

Marquette County, Upper Peninsula Michigan, United States

2018

Steven D.J. Baumann, Elisa J. Piispa, Eric Murray, and Sarah M. Hall

Publication Number: M-092018-1A

(Updated: August 19, 2019)

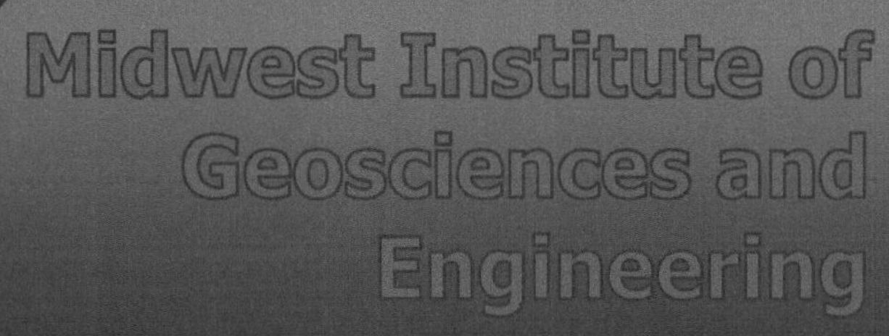

<u>**Geologic Units**</u>

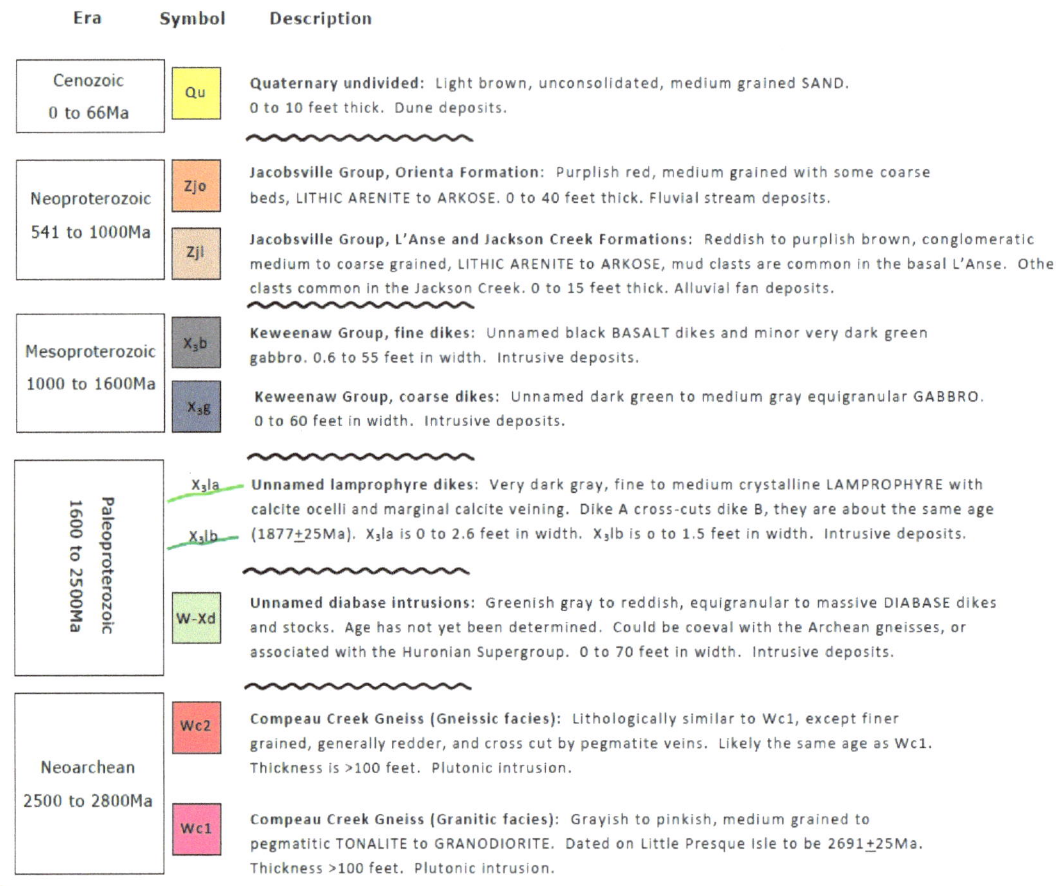

<u>**Geologic Map Symbols**</u>

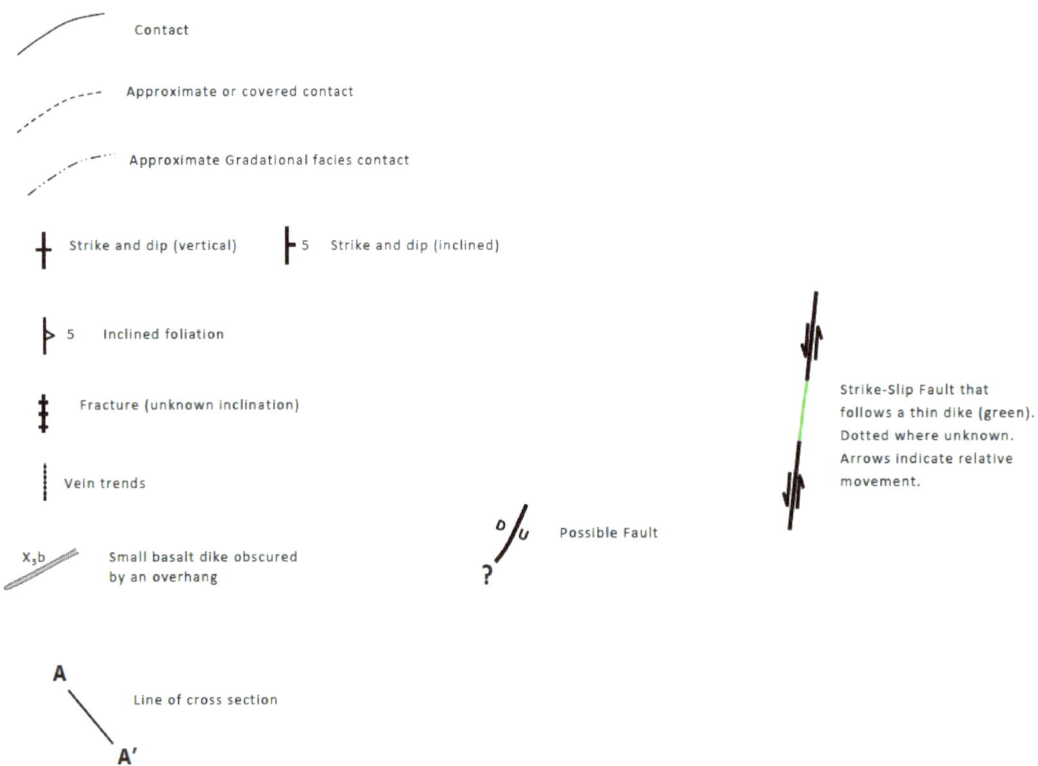

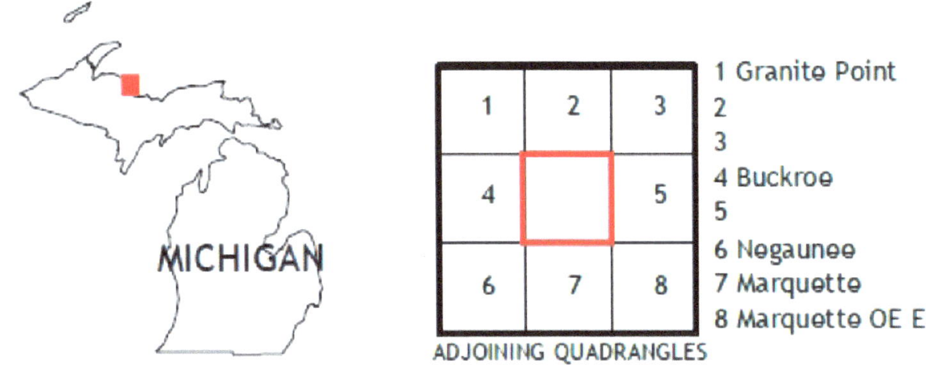

MARQUETTE NW, MI

2017

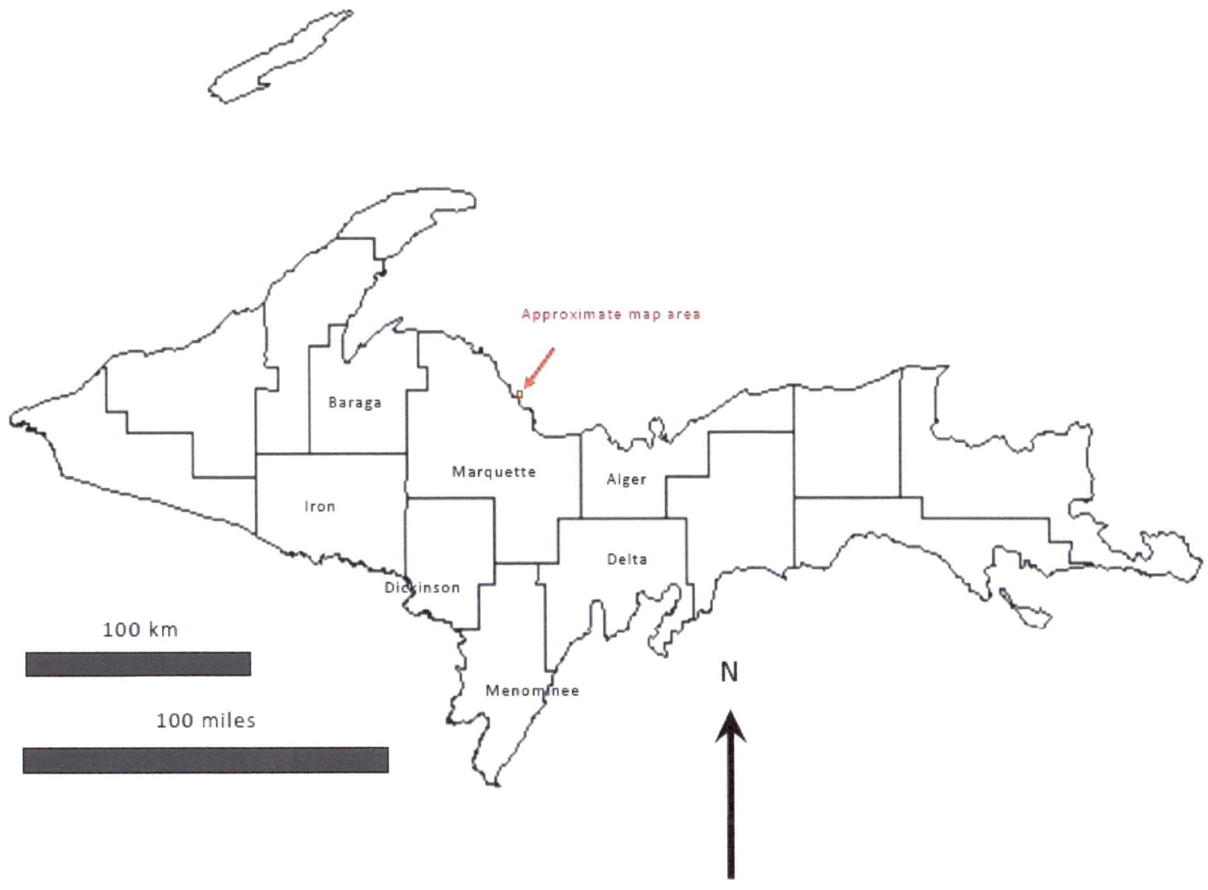

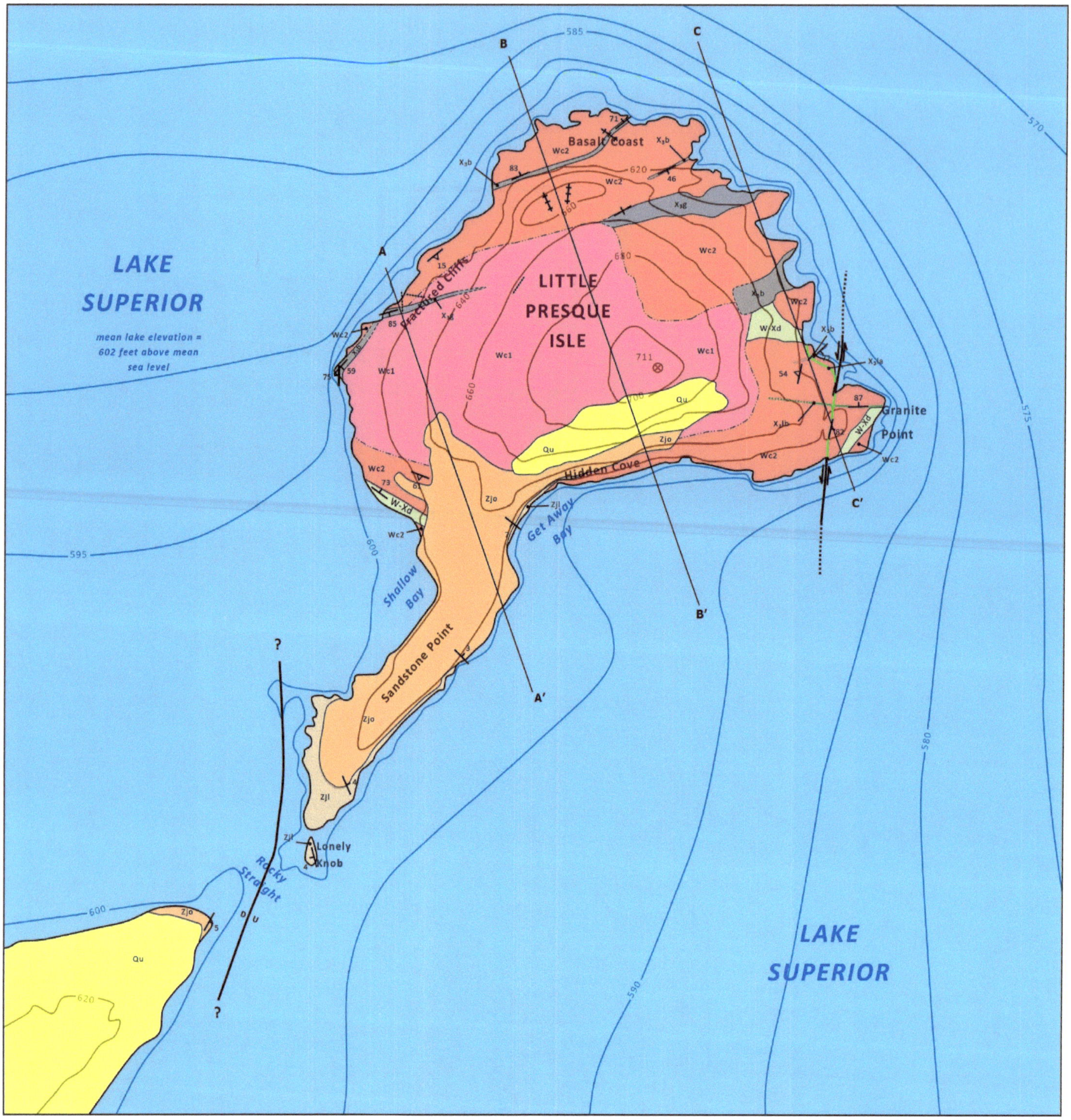

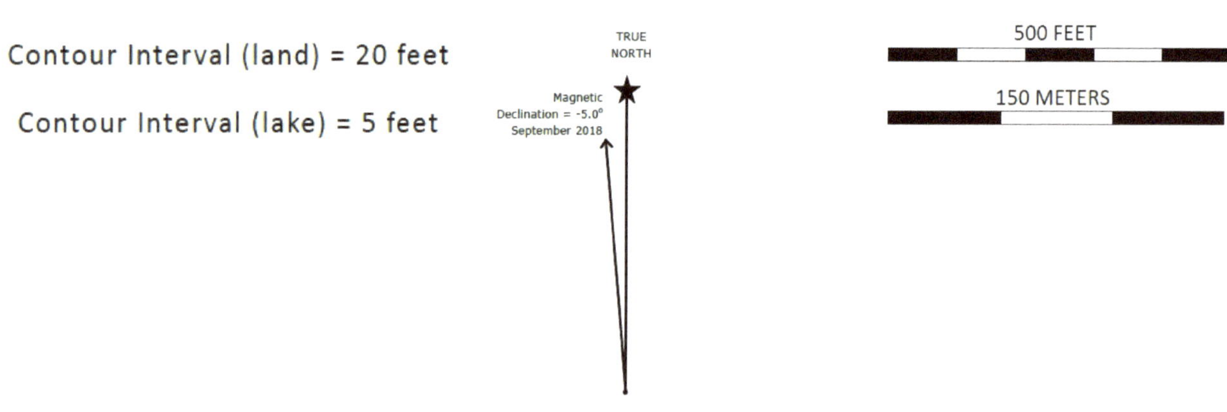

LAKE SUPERIOR
mean lake elevation = 602 feet above mean sea level
Basalt Coast
LITTLE PRESQUE ISLE
Fractured Cliffs
Hidden Cove
Get Away Bay
Granite Point
Shallow Bay
Sandstone Point
Rocky Straight
Lonely Knob
LAKE SUPERIOR
Contour Interval (land) = 20 feet
Contour Interval (lake) = 5 feet
TRUE NORTH
Magnetic Declination = -5.0°
September 2018
500 FEET
150 METERS

<u>**Cross Sections**</u>

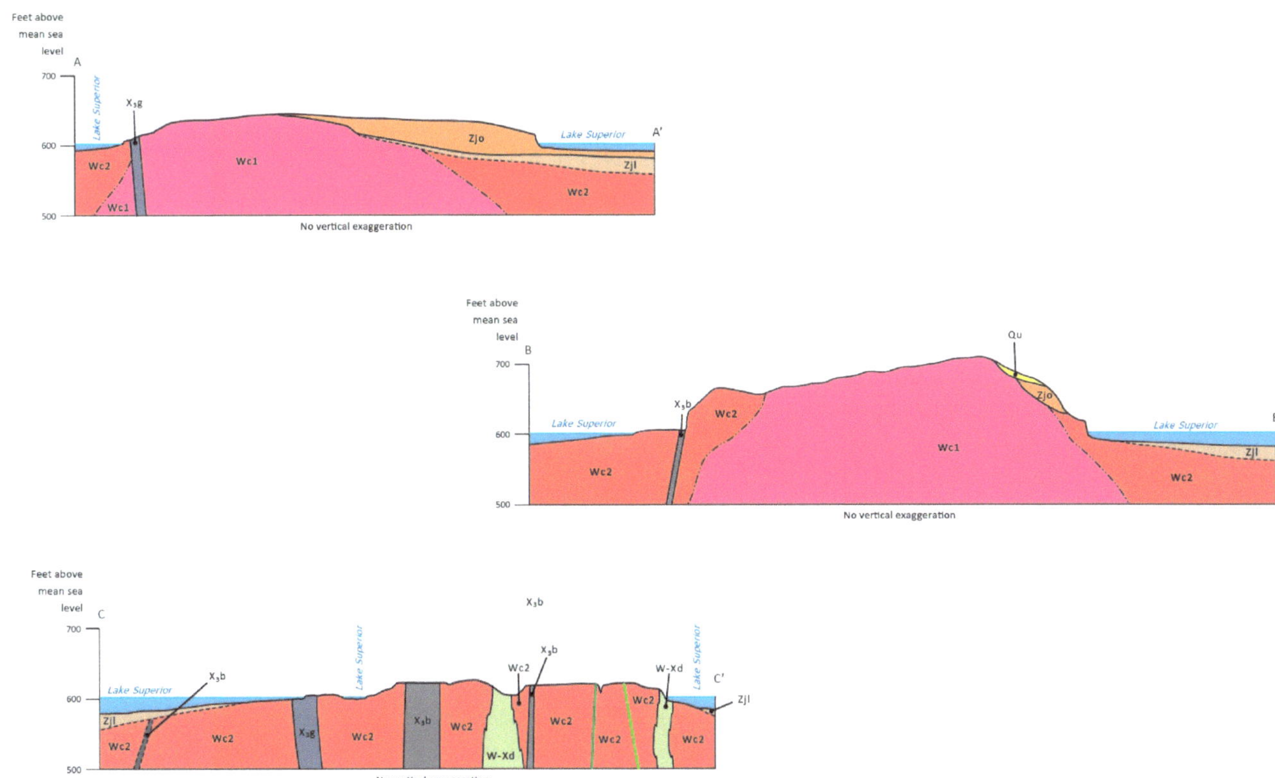

<u>Geology</u>

Method

Field mapping was conducted on August 11th 2018, September 3rd 2018, and August 11, 2019. A Brunton quadrant compass was used for structural measurements, an engineered tape measure was used for measuring dike widths, a GPS receiver was used to mark exact contacts, a field notebook was used to document activities, and photographs were taken with a digital camera.

Little Presque Isle (LPI) sits less than 500 feet from the mainland and can be walked to when the waves are calm. Lake Superior depth is 1 to 3.5 feet deep from the mainland to the LPI. Although the island is not very big, it is geologically complex.

Lithology and Structure

The host rock is the Archean Compeau Creek Gneiss, which is related to the rocks of the Wawa Terrane in Ontario (Craddock, et al. 2007). The gneiss has been successfully dated to 2691±25Ma. It shows two distinct and intertonguing facies. Mineralogically both facies are nearly identical but vary texturally. The first Wc1 is a mostly an undeformed, coarse grained, tonalite, and likely represents the core of a plutonic stock that was seeded deep underground. Its texture is largely granitic. The second gneissic facies Wc2, is much more deformed, showing a complex metamorphic history. It is unclear if Wc2 and Wc1 are coeval or if there was later metamorphism. Some foliations can be directly measured (see Figure 1) and the gneiss, in part, seems to dip towards the center of the island.

The diabase dikes of unit W-Xd are distinct. Dating was attempted but the rock did not yield usable zircons (Craddock, et al. 2007). They could have been emplaced right after the Compeau Creek Gneiss. Field relationships suggest a much later emplacement, perhaps during the Paleoproterozoic. We discovered another one on the west side of the island (see Figure 2). They are definitely older than the Keweenaw intrusions, which crosscut the diabase and likely older than the lamprophyre dikes (also crosscut by the Keweenaw).

The lamprophyre dikes, X_3la and X_3lb, were successfully dated to the time of the Penokean Orogeny at 1877 ± 25Ma (Craddock, et al. 2007). The lamprophyre dikes are likely near the same age, but not exactly. X_3la follows a strike-slip fault that offsets X_3lb, so X_3la is slightly younger (see Figures 3 and 4). Where dike X_3la turns west, may actually be a fault offset as well. However, it is under water (see geologic map).

Crosscutting all of these units are dark gabbros and basalts that are likely Mesoproterozoic in age and related to the Midcontinent Rift. Dating attempts on the island dikes were attempted but yielded no usable zircons (Craddock, et al. 2007). Based on lithology and trend, the dikes are likely between 1035Ma-1145Ma (Ernst and Buchan, 1993) and more likely closer to about 1140Ma (Piispa, 2015). This is based off the dated East Lake Superior lamprophyre (1143 ± 12Ma, Piispa, 2015), Marathon lamprophyre dikes (1145+15/-10Ma, Piispa, 2015), and the Abitibi diabase dike swarm (1140.6 ± 2.0Ma, Ernst and Buchan, 1993). During our field work we noticed a previously unmapped gabbro at Fractured Cliffs (see geologic map and Figure 5). This gabbro fits the local trends and may actually continue to the larger gabbro body along strike to the northeast side of the island. It is not known for sure if the two gabbros are the same intrusion since the top of the island is covered in trees and thick underbrush. It is unclear if the gabbro crosscuts the dominantly basaltic dikes or vice versa. Not all the Keweenaw dikes on the island are mapped. Some were too thin or under overhangs to be mapped. At the northeast side of the island is a 12-32" shallow dipping basalt dike that is hidden by an overhang (See Figure 6). It is the shallowest dipping dike on the island and has the best developed columnar jointing. The smallest known dike on the island is only 8" thick and is offset by a small strike slip fault with 18" of displacement. The displacement is too small to map but the fault trends N40W56SW (see Figure 7).

Jacobsville Group is the oldest sedimentary rock on the island and forms Sandstone Point as well as the ground underwater extending to the mainland. Three formations were recognized on LPI. However, only two are mapped. The basal Zjl is actually the basal valley fill L'Anse Formation. Right on top of the L'Anse (exposed on Lonely Knob) is a tongue of Jackson Creek Formation, which is mapped with the L'Anse, since it is generally less than 5 feet thick and is also conglomeratic. The overlying Zjo Formation contains rare pebbles usually only as lag conglomerate (Baumann, et al. 2017). There is a subtle syncline structure present on Sandstone Point (see Figure 8) that may indicate the presence of a small offshore fault (see geologic map). Others (Craddock, et al. 2007) have mapped the Jacobsville up to the top of the island. This is not correct. The high areas of the island only contain a thin cover of wind blown sands and organic debris, usually less than 3 feet thick (based on the fallen trees on the island). This is evident by the granitic and gneissic outcrops that are extensive on the high points of the island. There doesn't appear to be any Jacobsville above 660 feet mean sea level (MSL) on the island or the immediate surrounding area, although it likely did cover all of LPI during the Neoproterozoic.

There are no Paleozoic or Mesozoic deposits on LPI or in the local area. It is very doubtful that there were ever any Paleozoic deposits, except for maybe some Cambrian ones that have been eroded. Mesozoic deposits are even less likely.

Other than modern soils, there is a surprising lack of Quaternary deposits on the island. The immediate area of Lake Superior around LPI is bedrock made up of the Jacobsville and Archean rocks. There is no sandy bottom (see Figure 9). The only appreciable ones are represented by dune sands in the high area of Hidden Cove. The sand is a distinctly different color than the underlying Jacobsville, indicating it was not locally derived. Although glacial deposits and striations are lacking, there is a prominent cliff face carved around the island at about 625 ± 5 feet MSL (see cross sections). This could possibly be a high stand of Lake Superior. There is also large boulder talus (about the size of compact automobiles to large vans) around the Basalt Coast, covering a good portion of the exposed dike (see Figure 10).

<u>**Figures**</u>

FIGURE 1: GPS: 46.63849°N, 87.45743°W, Looking north, 6" blue pen for scale.

Showing the west-northwest dipping gneiss foliations on the southeast part of the island. Strike and dip is N12E54NW.

Photo taken on August 11, 2019. Taken by: Steven D.J. Baumann.

FIGURE 2: GPS: 46.63789°N, 87.46078°W, Looking north-northeast, 16 inch orange backpack for scale.

Outcrop of a newly discovered meta-diabase dike on the north end of Shallow Bay.

Photo taken on August 11, 2018. Taken by: Steven D.J. Baumann.

FIGURE 3: GPS: 46.63824°N, 87.45735°W, Looking north, 10" orange pouch for scale.

Showing the lamprophyre dike following the strike-slip fault X_3la. Yellow arrows show relative fault movement.

Photo taken on August 11, 2018. Taken by: Steven D.J. Baumann.

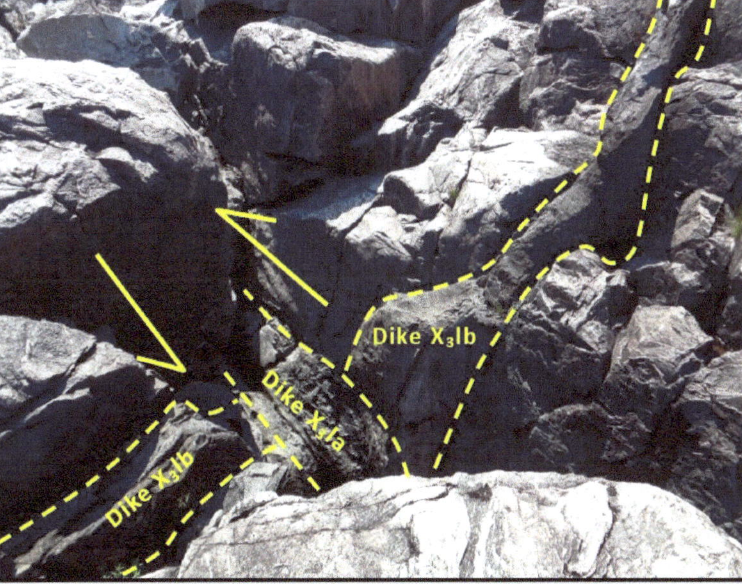

FIGURE 4: GPS: 46.63832°N, 87.45730°W, Looking southwest, dike X_3la is about 2.6' wide.

Showing the lamprophyre dike following the strike-slip fault X_3la. Yellow arrows show relative fault movement.

Photo taken on August 11, 2018. Taken by: Steven D.J. Baumann.

FIGURE 5: GPS: 46.63877°N, 87.46069°W, Looking west, 5'10" person for scale.

Showing the typical crosscutting relationships of the Keweenaw dikes and the Compeau Creek Gneiss.

Photo taken on August 11, 2018. Taken by: Steven D.J. Baumann.

FIGURE 6: GPS: 46.63956°N, 87.45873°W, Looking southwest.

The dike at the place where the phot was taken is about 32" thick, it narrows at the top of the hill.. Strike and dip is N63E46SE.

Photo taken on August 11, 2019. Taken by: Steven D.J. Baumann.

FIGURE 7: GPS: 46.63858°N, 87.45737°W, Looking northeast, 6" green and black pen for scale.

Showing the thinnest exposed basalt dike on the island. Strike and dip is N46E72SE. There is a small strike-slip fault that trends N40W 56SW.

Photo taken on August 11, 2019. Taken by: Steven D.J. Baumann.

FIGURE 8: GPS to the sandstone at center): 46.63744°N, 87.45979°W, Looking southwest, no scale.

Showing the contacts between the formations of the Jacobsville from Hidden Cove to Sandstone Point. The yellow arrow marks the approximate axis of the subtle syncline.

Photo taken on August 11, 2018. Taken by: Steven D.J. Baumann.

FIGURE 9: GPS: 46.63607°N, 87.46130°W, Looking north-northeast, 6 inch pen and people for scale.

The L'Anse Formation from Lonely Knob to Sandstone Point. This knob is in place. It is not an erratic.

Photo taken on August 11, 2018. Taken by: Steven D.J. Baumann.

FIGURE 10: GPS: 46.63979°N, 87.45893°W, Looking southwest, 10" orange pouch for scale.

Showing the talus blocks on top of the Keweenaw dike at Fractured Cliffs.

Photo taken on August 11, 2018. Taken by: Steven D.J. Baumann.

References

Baumann, S.D.J., Cory, A.B., Dylka, S.K., 2017, *Lithostratigraphic interpretation and redefinition of the sedimentary clastic assemblage (Bayfield Group, and Jacobsville Sandstone) in Michigan and Wisconsin, USA and Ontario, Canada*, Journal of Stratigraphy, V. 13, no. 3, pp. 163-181

Craddock, J.P., Anziano, J., Wirth, K., Vervoort, J.D., Singer, B., Zhang, X., 2007, *Structure, geochemistry and geochronology of a Penokean Lamprophyre Dike Swarm, Archean Wawa Terrane, Little Presque Isle, Michigan, USA*, Journal of Precambrian Research 157 (2007) 50-70, doi: 10.1016/j.precamres.2007.02.010

Ernst, R.E. and Buchan, K.L., 1993, *Paleomagnetism of the Abitibi dyke swarm, southern Superior Province, and implications for the Logan Loop*, Canadian Journal of Earth Science, v. 30, pp. 1886-1897

Piispa, E.J., 2015, *Precambrian geomagnetic field and geodynamics recorded by selected mafic dyke swarms in India and North America*, PhD Dissertation, Michigan Tech University

Topographic map, 1953, *Marquette NW Quadrangle, 7.5 Minute Series*, U.S. Department of the Interior, U.S. Geological Survey, AMS 3577 III NW-Series V862

Topographic map, 2017, *Marquette NW Quadrangle, Michigan-Marquette Co., 7.5 Minute Series*, U.S. Department of the Interior, U.S. Geological Survey, NSN 7643016371322

www.ingramcontent.com/pod-product-compliance
Lightning Source LLC
Chambersburg PA
CBHW042041110726

48006CB00002B/257